CARTOGRAPHIC MATERIALS

AMERICAN LIBRARY ASSOCIATION
MAP AND GEOGRAPHY ROUNDTABLE
Representative: Mary Larsgaard

ASSOCIATION OF CANADIAN MAP LIBRARIES
Representatives: Pierre Lepine
Joan Winearls

AUSTRALIAN MAP CURATORS CIRCLE
   and the
NATIONAL LIBRARY OF AUSTRALIA
Representative: Dorothy Prescott

BRITISH CARTOGRAPHIC SOCIETY
Representative: Roger Fairclough

BRITISH LIBRARY
and the
BRITISH COMMITTEE FOR MAP CATALOGUE SYSTEMS
Representatives: Yolande Hodson (Oct. 1979 – Dec. 1980)
Sarah Tyacke (Jan. 1981 –   )

LIBRARY OF CONGRESS
Representatives: David Carrington
John Schroeder
Ben Tucker

NATIONAL LIBRARY OF CANADA
Representative: Tom Delsey

NATIONAL MAP COLLECTION, PUBLIC ARCHIVES OF CANADA
Representatives: Vivien Cartmell
Velma Parker
Hugo L. P. Stibbe

NATIONAL LIBRARY OF NEW ZEALAND
Representative: Sheila Williams

NEW ZEALAND MAP KEEPERS CIRCLE
Representative: Phil Barton

SPECIAL LIBRARIES ASSOCIATION, GEOGRAPHY AND MAP
DIVISION
Representative: Mary Larsgaard

WESTERN ASSOCIATION OF MAP LIBRARIES
Representatives: Mary Larsgaard (Oct. 1979 – Nov. 1979)
Stanley Stevens (Nov. 1979 – Dec. 1980)
Myrna Fleming (Jan. 1981 –   )

CARTOGRAPHIC MATERIALS

A Manual of Interpretation for AACR2

Prepared by the Anglo-American
Cataloguing Committee for
Cartographic Materials

HUGO L. P. STIBBE
General Editor

VIVIEN CARTMELL and VELMA PARKER
Editors

CHICAGO   American Library Association   1982
OTTAWA   Canadian Library Association
LONDON   The Library Association

Published by

AMERICAN LIBRARY ASSOCIATION
50 East Huron Street
Chicago, Illinois 60611
ISBN 0-8389-0363-0

CANADIAN LIBRARY ASSOCIATION
151 Sparks Street
Ottawa, Ontario K1P 5E3
ISBN 0-88802-169-0

THE LIBRARY ASSOCIATION
7 Ridgmount Street
London WC1E 7AE
ISBN 0-85365-855-2

British Library Cataloguing in Publication Data
Anglo-American Cataloguing Committee for
  Cartographic Materials
  Cartographic materials: a manual of interpretation
  for AACR 2.
  1. Cataloguing of maps   2. Descriptive
  cataloguing—Rules
  I. Title   II. Stibbe, Hugo L. P.
  III. Cartmell, Vivien   IV. Parker, Velma
  025.3′46   Z695.6
  ISBN 0-85365-855-2

Canadian Cataloguing in Publication Data
Main entry under title:

Cartographic materials

Co-published by the Library Association.
ISBN 0-88802-169-0 (CLA)

1. Cataloging of maps—Handbooks, manuals, etc.
2. Descriptive cataloging—Rules—Handbooks, manuals,
etc.   I. Stibbe, Hugo L. P.   II. Cartmell, Vivien.
III. Parker, Velma.   IV. Anglo-American Cataloguing
Committee for Cartographic Materials.   V. American
Library Association.   VI. Canadian Library Association.
VII. Library Association.
Z695.6.C37   025.3′46   C82-090095-8

Library of Congress Cataloging in Publication Data
Main entry under title:

Cartographic materials.

  Includes bibliographical references and index.
  1. Cataloging of maps—Handbooks, manuals, etc.
2. Descriptive cataloging—Rules—Handbooks, manuals,
etc.   I. Stibbe, Hugo L. P.   II. Cartmell, Vivien.
III. Parker, Velma.   IV. Anglo-American Cataloguing
Committee for Cartographic Materials.
Z695.6.C37 1982   025.3′46   82-11519
ISBN 0-8389-0363-0

Printed in the United States of America

CONTENTS

Contents

Figures

PREFACE

The second edition of the *Anglo-American Cataloguing Rules (AACR2)* marked an advancement in cataloguing cartographic materials unequalled since the publication in 1945 of *The Classification and Cataloging of Maps and Atlases* by Samuel W. Boggs and Dorothy Cornwell Lewis. Every map librarian knows of this work which is still in use today. *AACR2*, in continuing the work of the first edition of 1967, has changes in presentation and content, but it is based on the same principles and underlying objectives as the earlier edition. The second edition has paid particular attention to developments in the machine processing of bibliographic records and standardization of description according to standards developed by the International Federation of Library Associations and Institutions (IFLA). In doing so, it has attained its greatest achievement of providing an integrated and standardized framework for the systematic description of all library materials. This achievement, perhaps more than any other, makes *AACR2* acceptable to all kinds of special libraries.

Dissatisfaction in the Anglo-American map collection community with the provisions for maps and atlases in the first edition of *Anglo-American Cataloging Rules (AACR1)* led to reactions ranging from a complete rejection to application with extensive modifications. In some instances where the rules were adopted, it was necessary to write extensive institutional internal policy manuals for interpretation and application.

With the emergence of map collections in libraries and of the map librarian as a recognizable category of special librarian in the 1950s, as well as an accelerated shift of map collections to the jurisdiction of libraries in the late 1960s and through the 1970s, the need for a uniform code for the cataloguing of cartographic materials was acutely felt. In addition, automation in libraries exerted an additional forceful incentive to standardize.

The Association of Canadian Map Libraries (ACML) was founded in 1967 and its National Union Catalogue Committee (NUC Committee) was established at the same time. This committee immediately recognized the need for standardization in both cataloguing and automation in order to attain their goal of a national union catalogue for cartographic materials. By 1969, the initial draft of a code, the *Canadian Cataloguing Rules for Maps*, had been compiled. The ACML intended to publish these rules as an alternative to the provisions for maps and atlases in *ACCR1*.

However, several developments changed the decision of the ACML. One was the development of the Universal Bibliographic Control (UBC) programme. As part of this programme, the Geography and Map Libraries Sub-Section (established in 1969, and now a section) was responsible for compiling the International Standard Bibliographic Description for Cartographic Materials (ISBD(CM)).[1] As well, there occurred the revision of *AACR1* with the stated objective of reconciling the British and North American texts, and the incorporation of the ISBD(G)[2] in the *Anglo-American Cataloguing Rules, second edition.*

After being revised to conform to ISBD(CM), the ACML draft rules were submitted in 1976 to the Canadian Committee on Cataloguing. In 1977, the ACML, the Geography and Map Division of the Library of Congress, and the United States Geological Survey jointly submitted comments on the draft chapter 3 *AACR2* to the Joint Steering Committee AACR (JSCAACR). Most of the recommended changes were included in *AACR2* chapter 3.

Now that the ACML was committed to the rules, its objective was to ensure the applicability of those rules to cartographic materials. The NUC Committee unanimously agreed that a manual, based on its original submission to JSCAACR, elucidating and where necessary amplifying the rules, was still required. To help ensure the wide acceptance of the use of *AACR2* by the map library community both in Canada and abroad, it was decided to solicit input for the manual from the Anglo-American community. Consequently, the ACML, on the initiative of Hugo Stibbe and through the sponsorship of the National Map Collection, Public Archives of Canada, convened a meeting in October 1979 in Ottawa, inviting cataloguing experts from the National Map Collection, the National Library of Canada, the Library of Congress, the British Library, and from British, American, and Canadian map library/curator associations. The group named itself the Anglo-American Cataloguing Committee for Cartographic Materials (AACCCM), and a secretariat was established at the National Map Collection, Public Archives of Canada. The committee was subsequently joined by representatives from Australia and New Zealand.

The document discussed at the 1979 Ottawa meeting was based on the ACML submission to JSCAACR which had been reworked to conform to *AACR2*. In Great Britain, a committee of map librarians from national and other major map collections was organized in 1979 to consider the application of automation to cataloguing cartographic materials under the auspices of the British Library. The group, named the British Machine Readable Records Maps Steering Committee,[3] formed a special Working Party to consider the proposed manual and to make recommendations to the AACCCM.

After the October 1979 meeting, another draft was prepared and issued to the Committee. This document was finalized at the second AACCCM meeting held in Washington, D.C., in April-May 1981, and it has resulted in this manual which is designed to accompany *AACR2*.

Since the publication of *AACR2* in 1978, the *ISBD(A)*[4] and the *Bibliographic Descrip-*

---

1. International Federation of Library Associations and Institutions. *ISBD(CM) : International Standard Bibliographic Description for Cartographic Materials.* London : IFLA International Office for UBC, 1977.

2. International Federation of Library Associations and Institutions. *ISBD(G) : General International Standard Bibliographic Description : Annotated Text.* London : IFLA International Office for UBC, 1977.

3. Renamed the British Committee for Map Catalogue Systems in 1981.

4. International Federation of Library Associations and Institutions. *ISBD(A) : International Standard Bibliographic Description for Older Monographic Publications (Antiquarian).* London : IFLA International Office for UBC, 1980.

tion of Rare Books,[5] both of which deal with older materials, have been issued. The latter document is based on *AACR2* and *ISBD(A)*. Having considered the applicability of both these documents to cartographic materials, the AACCCM decided against incorporating more detailed guidance for cataloguing such material in the manual at this time. It felt there had been insufficient time to thoroughly study the documents and to develop guidelines for the comprehensive treatment of early materials for the manual.

ACKNOWLEDGEMENTS

The task of preparing the manual has been challenging and often arduous for the many individuals who have been engaged in it. Thanks are extended to the organizations, committees, and individuals in several countries who have devoted time and trouble to assist in this enterprise: their names are on the following pages. Barbara Farrell deserves special thanks for designing and drafting the majority of the illustrations; and similarly Barbara Christie for supplying most of the in-text cartographic examples.

Finally, grateful acknowledgement is made to the National Map Collection, Public Archives of Canada, and the Library of Congress, Geography and Map Division for the financial and other supportive resources with which they sustained the project.

HUGO L. P. STIBBE

CANADA

Members of the National Union Catalogue Committee of the Association of Canadian Map Libraries since 1978 when the Committee decided to produce an international manual:

> Vivien Cartmell
> Kate Donkin, Chairman
> Lorraine Dubreuil
> Barbara Farrell
> Pierre Lepine
> Velma Parker
> Hugo L. P. Stibbe
> Yves Tessier
> Grace Welch
> Joan Winearls

NEW ZEALAND

Resource person: Tony Ralls (National Library of New Zealand)

---

5. United States. Library of Congress. Office for Descriptive Cataloging Policy. *Bibliographic Description of Rare Books.* Washington : Library of Congress, 1981.

Preface

UNITED KINGDOM

Thanks are due the following members and invited observers of the British Committee for Map Catalogue Systems, and its Working Party.

P. Barber (Department of Manuscripts, British Library)
Margaret Brennand (Public Record Office)
R. Carpenter (Bibliographic Standards Office, Bibliographic Services Division, British Library)
R. A. Christophers (Catalogue Systems Branch, Reference Division, British Library)
P. K. Clark (Ministry of Defence Map Library)
Moira Courtman (University of London, University Library)
R. Davies (National Library of Wales)
R. H. Fairclough (Cambridge University Library)
Betty Fathers (Bodleian Library)
I. P. Gibb (Reference Division, British Library)
F. Herbert (Royal Geographical Society)
Yolande Hodson (Map Library, British Library)
P. Milne (National Library of Scotland)
D. Moore (National Library of Wales)
C. Terrell (National Maritime Museum)
Sarah Tyacke (Map Library, British Library)
Helen Wallis (Map Library, British Library)
G. Webster (Ministry of Defence Map Library)
Margaret Wilkes (National Library of Scotland)

UNITED STATES

Resource persons: Barbara Christie (Library of Congress)
Richard R. Fox (Library of Congress)
Robert Karrow (Newberry Library)
Dorothy McGarry (University of California, Los Angeles)
Minnie A. Modelsky (Library of Congress)

GENERAL INTRODUCTION

This manual is designed for use in conjunction with the second edition of the *Anglo-American Cataloguing Rules (AACR2)* and is not intended to stand on its own. Its purpose is to elucidate *AACR2* by means of applications, policies, and examples. The manual is directed to libraries that have a special need for detailed cataloguing of their cartographic material holdings, to national agencies that prepare the national bibliographic records for cartographic materials, and to cataloguers who, having no specialized knowledge of cartographic materials, need further guidance than that given in *AACR2*. The objectives established at the first meeting of the AACCCM may be briefly stated as follows:

1) To support the concepts, general principles, and much of the specific content of *AACR2*
2) To provide maximum conformity with *AACR2*
3) To provide as a general principle maximum uniformity of description and access to information, regardless of format, in order to meet the reference requirements of general libraries
4) To provide expansions and interpretations of *AACR2*, particularly chapter 3, in order to achieve standardization, consistency, and precision in cataloguing cartographic materials
5) To resolve some basic problems in cataloguing cartographic materials such as:
   a) the inconsistencies in the arrangement of bibliographic information on maps and the scattering of bibliographic elements
   b) the frequent omission of key bibliographic elements, e.g., date
   c) the differences in concepts and definitions of terms between cartographers, map librarians, and librarians dealing with other kinds of library materials.

With regard to the above objectives, this is a manual of interpretation within the framework of *AACR2*, specially compiled to facilitate the achievement of uniformity of description and entry and also to ensure maximum compatibility of bibliographic description between cartographic and other materials. Most of the manual deals with *AACR2* Part I (Description), but this does not imply that Part II (Headings, Uniform Titles, and References) is any less significant.

The second part of *AACR2* is applicable as written except for additional guidance in

Appendix A of the manual on rules 21.1A1 and 21.1B2 from chapter 21 (Choice of Access Points). This demonstrates a reconciliation between the North American and British texts of the first edition of *AACR*. In the introductory note to chapter 11 of the latter text, it is pointed out that the distinctive characteristics of maps require special treatment in a catalogue entry, and that most map libraries using the British text make the main entry under the name of the area depicted on the map. However, developments in cataloguing theory, the automation of library catalogues, and data retrieval have made obsolete the precepts on which the above statements were based.

The second edition of *AACR* achieves compatibility between entries made for cartographic material forming part of a general catalogue of library materials and also for entries compiled for a separate map catalogue. Neither this manual nor *AACR2* provides rules for map catalogues which make their main entries under geographical area. The concept of area access is more akin to subject analysis with separate access created for both the geographic area and the thematic subject (geology, climatology, etc.). The *Anglo-American Cataloguing Rules* do not provide guidance on subject access and neither does this manual. It is recognized that map libraries will continue to require geographical area access, and the AACCCM may recommend standard procedures for such access in the future.

Organization of the manual

Part I of *AACR2* deals with the provision of information describing the item being catalogued, and Part II deals with the determination and establishment of access points in the catalogue, under which the descriptive information is to be presented to users. The rules of Part I (Description) were considered most in need of clarification in order to be applied to cartographic materials. The difference in the applicability to cartographic materials of Parts I, II, and the Appendices has affected the organization of the manual in the following ways:

1) All rules of Part I (Description) have been brought together in one sequence. The rules of Part I considered inapplicable to cartographic materials have been omitted.

2) All of Part II (Headings, Uniform Titles, and References) is applicable but has been omitted except for the additional guidance on rules 21.1A1 and 21.1B2 which appears in Appendix A of this manual.

3) All of Appendix A (Capitalization) of *AACR2* is applicable but omitted. The relevant parts of Appendices B (Abbreviations), C (Numerals), and D (Glossary) are included.

Consequently, the manual must be used in conjunction with *AACR2*. This organization, in addition to eliminating much duplication of text, retains compatibility with *AACR2*.

Rule numbering

The rule numbers of the manual generally follow the system used in *AACR2*, although the chapter numbers have been eliminated. Most rules in the manual have been subsumed under those of chapter 1 of *AACR2* except chapter 13 (Analysis), which retains the designation 13 in the manual. To aid reference to *AACR2*, the *AACR2* rule number is quoted in parentheses following the rule.

Rules, applications, and policies

Most rules in the manual as well as some of their accompanying examples have been quoted verbatim from *AACR2*. Some rules have been modified, without altering their meaning, in order to make them intelligible when applied to cartographic materials. In a limited number of cases, rules have been corrected, or an amendment substituted for the incorrect rule.[1] These rules are indicated by the abbreviation ''mod.'' following the *AACR2* number. The guidance matter is divided into two categories: applications and policies.

1) *Applications* interpret rules and supply guidance for their use.
2) *Policies* occur only in rules and applications which include options. The policies noted under this heading are those of national agencies responsible for their national bibliographic record for cartographic materials. Missing national policy statements indicate that no statement was received at the time the manual went to press and does not necessarily mean that the agency does not have a policy with respect to the option offered.

Concordance

A concordance from *AACR2* rules to those of the manual immediately precedes the index. The concordance from the rules of the manual to those of *AACR2* appears within the text where the *AACR* number is in parentheses adjacent to the quoted rule.

Examples

The General Introduction of *AACR2* states, ''The examples throughout these rules are illustrative and not prescriptive. That is, they are intended to illuminate the provisions of the rule to which they are attached, rather than to extend those provisions.'' The manual also follows this policy. Where available, cartographic examples have been substituted for non-cartographic examples. The examples in Appendix G are provided by the national agencies who contributed to the manual. These full descriptive entries, which lack headings or other added entries, illuminate the use of *AACR2* and the manual by the contributing agency.

Appendices, glossary, and index

The appendices of the manual fall into the following categories:

1) Those with instructions that have the same force as the rules themselves and should be applied consistently, i.e., Appendices A, Guidelines for Choice of

---

1. Such amendments are submitted for approval to the reconstituted Joint Steering Committee for the revision of *AACR2* (2JSCAACR). Users of this manual who wish to keep it up to date are advised to subscribe to the publication in their country where the decisions on, and the applications of, rule revisions approved by 2JSCAACR are documented.

Access Points; H, Abbreviations; and J, Numerals. Appendices H and J have been compiled from *AACR2* appendices B and C respectively, and include only those rules relevant to cartographic materials.

2) Those providing further guidance on specific problems pertaining to cartographic materials, i.e., Appendices B, Scale; C, Date; D, Series; E, Treatment of Map Series; F, Geographic Atlases.

3) Appendix G comprises examples of full descriptive entries.

The glossary consists of terms from various sources, including those from the glossary of *AACR2* that are relevant to cataloguing cartographic materials.

The index is based on that of *AACR2* and follows its arrangement.

DESCRIPTION

Alternatives and options

0.1 Some rules are designated as *alternative rules* or as *optional additions,* and some other rules or parts of rules are introduced by the word *optionally*. These provisions arise from the recognition that different solutions to a problem and differing levels of detail and specificity are appropriate in different contexts. Some alternatives and options should be decided as a matter of cataloguing policy for a particular catalogue or bibliographic agency and should therefore be exercised either always or never. Other alternatives and options should be exercised case by case. It is recommended that all cataloguing agencies distinguish between these two types of options and keep a record of their policy decisions and of the circumstances in which a particular option may be applied. (0.7)

0.2 The word *prominently* (used in such phrases as *prominently named* and *stated prominently*) means that a statement to which it applies must be a formal statement found in one of the prescribed sources of information (see 0B) for areas 1 and 2 for the class of material to which the item being catalogued belongs. (0.8)

0.3 The necessity for judgement and interpretation by the cataloguer is recognized in these rules. Such judgement and interpretation may be based on the requirements of a particular catalogue or upon the use of the items being catalogued. The need for judgement is indicated in these rules by words and phrases such as *if appropriate, important,* and *if necessary*. These indicate recognition of the fact that uniform legislation for all types and sizes of catalogues is neither possible nor desirable, and encourages the application of individual judgement based on specific local knowledge. This statement in no way contradicts the value of standardization. Such judgements must be applied consistently within a particular context and must be recorded by the cataloguing agency. (0.9)

0.4 It is a cardinal principle of the use of *AACR2* Part I that description of a physical item should be based in the first instance on the chapter dealing with the class of materials to which that item belongs. For example, a printed monograph in microform should be described as a microform (using the rules in *AACR2* chapter 11). There will be need in many instances to consult the chapter dealing with the original form of the item, especially when constructing notes. So, using the same example, the chapter dealing with printed books (*AACR2* chapter 2) will be used to supplement chapter 11.

In short, the starting point for description is the physical form of the item in hand, not the original or any previous form in which the work has been published.

In describing serials, *AACR2* chapter 12 should be consulted in conjunction with the chapter dealing with the physical form in which the serial is published. So, in describing serial motion pictures, both chapters 12 and 7 should be used. (0.24)

■APPLICATION   In describing cartographic material, this manual should be consulted in conjunction with the chapter dealing with the physical form in which the cartographic material appears. It may appear in the form of material covered by at least the following chapters in *AACR2*: 2 (Books, Pamphlets, and Printed Sheets), 4 (Manuscripts), 8 (Graphic Materials), 10 (Three-Dimensional Artefacts and Realia), 11 (Microforms), 12 (Serials).

0.5   An innovation of the ISBD(G) is the introduction of an area for details that are special to a particular class of material or type of publication. This area (area 3) is used in these rules for cartographic materials and for serials (see *AACR2* chapter 12).[1] In describing a serial that consists of cartographic materials (e.g., a map series), area 3 may be repeated. In such case, give the area 3 details relating to cartographic materials before those relating to the serial. (0.25 mod.)

0.6   The measurements prescribed are not all metric. They are the normal measurements used at this time in libraries in Canada, the United Kingdom, and the United States. Where no predominant system of measurement exists, metric measurements have been used. Metric measurements may be substituted for the nonmetric measurements when:

either   a)   in the course of time a metric measurement becomes the normal measurement for the materials in question

or   b)   the rules are being used in a country where only metric measurements are used. (0.28 mod.)

0   **GENERAL RULES** (3.0)

0A   Scope

The rules in this manual cover the description of cartographic materials of all kinds. Cartographic materials include all materials that represent, in whole or in part, the earth or any celestial body. These include two- and three-dimensional maps and plans (including maps of imaginary places); aeronautical, navigational, and celestial charts; atlases; globes; block diagrams; sections; aerial photographs with a cartographic purpose; bird's-eye views (map views); etc. They do not cover in detail the description of early or manuscript cartographic materials, though the use of an additional term in the physical description (see 5B) and the use of the specific instructions in *AACR2* chapter 4 will furnish a sufficiently detailed description for the general library catalogue. (3.0A)

---

1. This paragraph mentions the use of *AACR2* chapter 12 in connection with cartographic material and gives ''map series'' as an example. While cartographic serials exist, they are rare, and the category of materials called ''map series'' is not a serial. See the Glossary for the necessary distinction.

0A1   Organization of the description. The description is divided into the following areas:

> Title and statement of responsibility
> Edition
> Material (or type of publication) specific details
> Publication, distribution, etc.
> Physical description
> Series
> Note
> Standard number and terms of availability

Each of these areas is divided into a number of elements as set out in the rules.

(1.0B)

0B   Sources of information   (3.0B)

0B1   This manual contains specifications of the chief sources of information for carto-graphic materials. A source of information may be unitary in nature (e.g., a title page) or may be collective. Prefer information found in that chief source to information found elsewhere. For each area of the description one or more sources of information are prescribed. Enclose in square brackets information taken from outside the prescribed source or sources.   (1.0A1 mod.)

0B2   Items lacking a chief source of information. If no part of the item supplies data that can be used as the basis of the description, take the necessary information from any available source, whether this be a reference work or the content of the item itself. This technique may be necessary for printed works, the title pages of which are lost; collec-tions of pamphlets or other minor material assembled by the library or by a previous owner and which are to be catalogued as a single item, etc. In all such cases give in a note the reason for and/or source of the supplied data.   (1.0A2 mod.)

0B3   Chief source of information. The chief source of information (in order of pref-erence) is:

a)   the cartographic item itself; when an item is in a number of physical parts, treat all the parts (including a title sheet) as the cartographic item itself
b)   container (portfolio, cover, envelope, etc.) or case, the cradle and stand of a globe, etc.

If the information is not available from the chief source, take it from any accompany-ing printed material (pamphlets, brochures, etc.)   (3.0B2)
See Figure 1.

■APPLICATION   The chief source of information is the cartographic item itself with a container (portfolio, cover, envelope, etc.) or case, the cradle and stand of a globe, etc., issued by the publisher or manufacturer of the item. When an item is in a number of physical parts, treat all the parts (including a title sheet) as the cartographic item itself

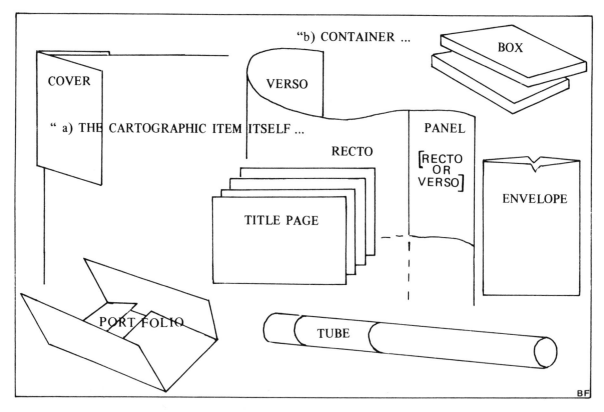

Figure 1   0B3   Chief source of information

(see Figure 1). Prefer information found on the cartographic item itself (including any permanently affixed labels) to information found on a container.

Indicate the source of information in a note if taken from other than a prescribed source of information.

0B4   Chief source of information for an atlas. The chief source of information for printed atlases is the title page or, if there is no title page, the source from within the publication that is used as a substitute for it. For printed atlases published without a title page, or without a title page applying to the whole work, use the part of the item supplying the most complete information, whether this be the cover (excluding a separate book jacket), half title page, caption, colophon, running title, or other part. Specify the part used as a title page substitute[2] in a note (see 7B3). If no part of the item supplies data that can be used as the basis of the description, take the necessary information from any available source. If information traditionally given on the title page is given on two

---

2. Hereafter in the rules pertaining to atlases the term *title page* is used to include any substitute.

facing pages or on pages on successive leaves, with or without repetition, treat those pages as the title page. (2.0B1 mod.)

0B5   Chief source of information for early printed atlases. If the early atlas has a title page, use it as the chief source of information. If it has no title page, use the following sources (in this order of preference):

> half title
> caption
> colophon
> cover
> running title
> incipit *or* explicit
> privilege *or* imprimatur
> other sources

Make a note indicating the source used if the item has no title page (see 7B3).

(2.13 mod.)

■APPLICATION   If there is no title page and one of the above sources of information is used, do not enclose the title in square brackets.

0B6   Prescribed sources of information. The prescribed source(s) of information for each area of the description of cartographic materials is set out below. Enclose information taken from outside the prescribed source(s) in square brackets.

AREA	PRESCRIBED SOURCES OF INFORMATION
Title and statement of responsibility	Chief source of information
Edition	Chief source of information, accompanying printed material
Mathematical data	Chief source of information, accompanying printed material
Publication, distribution, etc.	Chief source of information, accompanying printed material
Physical description	Any source
Series	Chief source of information, accompanying printed material
Note	Any source
Standard number and terms of availability	Any source

(3.0B3)

0B7   Prescribed sources of information for atlases. The prescribed source(s) of information for each area of the description of printed atlases is set out below. Enclose information taken from outside the prescribed source(s) in square brackets.

AREA	PRESCRIBED SOURCES OF INFORMATION
Title and statement of responsibility	Title page
Edition	Title page, other preliminaries, and colophon
Mathematical data	The whole publication
Publication, distribution, etc.	Title page, other preliminaries, and colophon
Physical description	The whole publication
Series	The whole publication
Note	Any source
Standard number and terms of availability	Any source

(2.0B2 mod.)

■APPLICATION   In all cases in which data for the title and statement of responsibility area, the edition area, and the publication, distribution, etc., area are taken from other than the prescribed sources, make a note to indicate the source of the data.

0B8   Prescribed sources of information for manuscripts. The prescribed source(s) of information for each area of the description of manuscript texts is set out below. Enclose information taken from outside the prescribed source(s) in square brackets.

AREA	PRESCRIBED SOURCES OF INFORMATION
Title and statement of responsibility	Chief source of information and manuscript or published copies
Mathematical data	Chief source of information and accompanying material
Date	Chief source of information and manuscript or published copies
Physical description	Any source
Note	Any source

(4.0B2 mod.)

0C   Punctuation

For the prescribed punctuation of elements, see the following rules.

(3.0C)

0C1   Precede each area, other than the first area, or each occurrence of a note or standard number, etc., area by a full stop, space, dash, space (. —) unless the area begins a new paragraph.

Precede or enclose each occurrence of an element of an area with standard punctuation prescribed at the head of each area of this manual, i.e., 1 Title, etc. area; 2 Edition area; etc.

Precede each mark of prescribed punctuation by a space and follow it by a space except for the comma, full stop, hyphen (see *AACR2* 12.3A1), and opening and closing parentheses and square brackets. The comma, full stop, hyphen, and closing parenthesis and square bracket are not preceded by a space; the hyphen and the opening parenthesis and square bracket are not followed by a space.

Precede the first element of each area, other than the first element of the first area or the first element of an area beginning a new paragraph, by a full stop, space, dash, space. When that element is not present in a description, precede the first element that is present by a full stop, space, dash, space instead of the prescribed preceding punctuation for that element.

Indicate an interpolation (i.e., data taken from outside the prescribed source(s) of information) by enclosing it in square brackets. Indicate a conjectural interpolation by adding a question mark within the square brackets. Indicate the omission of part of an element by the mark of omission (. . .). Precede and follow the mark of omission by a space. Omit any area or element that does not apply in describing an individual item; also omit its prescribed preceding or enclosing punctuation. Do not indicate the omission of an area or element by the mark of omission.

When adjacent elements within one area are to be enclosed in square brackets, enclose them in one set of square brackets unless one of the elements is a general material designation, which is always enclosed in its own set of brackets.   (1.0C mod.)

> Urbanismo español en América [GMD] : [exposición itinerante por Hispanoamérica / organizada por el] Instituto de Cultura Hispánica ; [proyectada y dirigida por] Javier Aguilera Rojas, Joaquín Ibañez Montoya, Luis J. Moreno Rexach

> *but*   [London : Phipps, 1870]

When adjacent elements are in different areas, enclose each element in a set of square brackets.

> [2nd ed.]. — [London] : Thomsons, 1973

When an element ends with an abbreviation followed by a full stop or ends with the mark of omission and the punctuation following that element either is or begins with a full stop, omit the full stop that constitutes or begins the prescribed punctuation.

> 261 p. ; 24 cm. — (Canadian Ethnic Studies Association series ; v. 4)
> *not*   261 p. ; 24 cm.. — (Canadian Ethnic Studies Association series ; v. 4)

When punctuation occurring within or at the end of an element is retained, give it with normal spacing. Prescribed punctuation is always added, even though double punctuation may result.   (1.0C)

> Quo vadis? : a narrative from the time of Nero

Summary of normal sequence and punctuation

1. Multiparagraph format
    Title [general material designation] = parallel title : other title information / first statement of responsibility ; second statement of responsibility ; third statement of responsibility. — Edition statement / first statement of responsibility ; second statement of responsibility,

second edition statement / first statement of responsibility ; second statement of responsibility. — Scale ; projection (coordinates ; equinox). — First place of publication ; second place of publication : publisher, date (Place of manufacture : manufacturer, date)

Number special material designation : other physical details ; size + accompanying material (physical details of accompanying material). — (Series title = parallel title : other title information / first statement of responsibility ; second statement of responsibility, ISSN ; numbering within series. Title of subseries)

Introductory wording of note: First note
Second note
Third note
etc.
Standard number (qualification) = Key-title : terms of availability (qualification)

2.   Single paragraph format
Title ₁general material designation₁ = parallel title : other title information / first statement of responsibility ; second statement of responsibility ; third statement of responsibility. — Edition statement / first statement of responsibility ; second statement of responsibility, second edition statement / first statement of responsibility ; second statement of responsibility. — Scale ; projection (coordinates ; equinox). — First place of publication ; second place of publication : publisher, date (Place of manufacture : manufacturer, date). — Number special material designation : other physical details ; size + accompanying material (physical details of accompanying material). — (Series title = parallel title : other title information / first statement of responsibility ; second statement of responsibility, ISSN ; numbering within series. Title of subseries). — Introductory wording of note: First note. — Second note. — Third note. — etc. — Standard number (qualification) = Key-title : terms of availability (qualification)

3.   Multilevel description
The punctuation pattern is as illustrated above with the exception of instances where sheet designations are present. When the sheet designation is present, the second level pattern is as follows:

Sheet designation: title / statement of responsibility . . .

■APPLICATION
Early cartographic items. If desired, record all of the punctuation that is found in the sources of information. When this punctuation is recorded, always give the prescribed punctuation as well, even if this results in double punctuation. When the original punctuation mark and the prescribed punctuation mark are the same, give only the prescribed punctuation mark. In the latter case, *optionally* give both the original and the prescribed punctuation marks.

Indicate omissions by the cataloguer by the mark of omission (. . .). Indicate lacunae in the source affecting part of an element by the mark of omission enclosed in square brackets.

12

When an entire element or area is not present in the source, hence not transcribed, omit the corresponding prescribed punctuation from the transcription. Do not indicate the omission of an entire element or area by the mark of omission.

When omitting information from the source that is not considered part of any area (i.e., pious invocations, etc.), do not use the mark of omission.

Record the mark of omission with a space on both sides of it. If the mark of omission precedes a full stop, omit the full stop, even if it is prescribed punctuation.

ɟtitleɟ / ɟstatement of responsibilityɟ . . . ɟtitleɟ / ɟstatement of responsibilityɟ

ɟtitleɟ . . . — ɟedition statementɟ

Do not transcribe a mark of punctuation that precedes the mark of omission unless it is a mark of abbreviation or of final punctuation (e.g., a full stop or a question mark).

☐ POLICIES

British Library
The "optionally" provision is not applied.

Library of Congress
The "optionally" provision is not applied.

National Library of Australia
The "optionally" provision is not applied.

National Map Collection, PAC
The "optionally" provision is not applied.

0D Levels of detail in the description

0Da This rule contains a specification of three levels of description. Each of these levels is to be considered as a minimum in that, when appropriate, further information may be added to the required set of data. The three levels of description allow libraries flexibility in their cataloguing policy, because they prescribe an entry that is in conformity with bibliographic standards and yet allow some materials to be described in more detail than others. Libraries may choose to use the three levels of description:

either i) by choosing a level of description for all items catalogued in that library
or ii) by drawing up guidelines for the use of all three levels in one catalogue depending on the type of item being described.

This standardization at three levels of description will help in achieving uniformity of cataloguing, and it is recommended that each record in a machine system carry an indication of the level at which the item has been described. (0.29)

0Db The elements of description provided in the rules constitute a maximum set of information. This rule sets out three recommended levels of description containing those elements that must be given as a minimum by libraries and other cataloguing agencies choosing the level of description. Base the choice of a level of description on the purpose of the catalogue or catalogues for which the entry is constructed. Include this minimum set of elements for all items catalogued at the chosen level when the elements are applicable to the item being described and when, in the case of *optional additions*, the library has chosen to include an optional element. If the rules in *AACR2* Part I specify other pieces of information in place of any of the elements set out below, include those other pieces of information. Consult individual rules for the content of elements to be included. (1.0D)

0Dc Additionally, in a first-level description include the scale in the mathematical data area. In a second-level description, include all the data specified in area 3.

(3.0D)

0D1 First level of description. For the first level of description, include at least the elements set out in this schematic illustration:

> Title proper / first statement of responsibility, if different from main entry heading in form or number or if there is no main entry heading. — Edition statement. — Material (or type of publication) specific details. — First publisher, etc., date of publication, etc. — Extent of item. — Note(s). — Standard number.

See 1B, 1F, 2B, 3, 4D, 4F, 5B, 7, 8B. (1.0D1)

■APPLICATION   Include scale only in material specific details, and series statements in the first-level description.

0D2 Second level of description. For the second level of description, include at least the elements set out in this schematic illustration:

> Title proper ₍general material designation₎ = parallel title : other title information / first statement of responsibility ; each subsequent statement of responsibility. — Edition statement / first statement of responsibility relating to the edition. — Material (or type of publication) specific details. — First place of publication, etc. : first publisher, etc., date of publication, etc. — Extent of item : other physical details ; dimensions. — (Title proper of series / statement of responsibility relating to series, ISSN of series ; numbering within the series. Title of subseries, ISSN of subseries ; numbering within subseries). — Note(s). — Standard number.

(1.0D2)

0D3 Third level of description. For the third level of description, include all elements set out in the following rules that are applicable to the item being described.

(1.0D3)

0E Language and script of the description (3.0E)

In the following areas, give information transcribed from the item itself in the language and script (wherever practicable) in which it appears there:

> Title and statement of responsibility
> Edition
> Publication, distribution, etc.
> Series

Replace symbols or other matter that cannot be reproduced by the typographical facilities available with a cataloguer's description in square brackets. Make an explanatory note if necessary. (Cf. 1B1, 1F9, 2B2.)

In general, give interpolations into these areas in the language and script of the other data in the area. Exceptions to this are:

1) prescribed interpolations and abbreviations
2) general material designations (see 1C)
3) other forms of the place of publication (see 4C2)
4) statements of function of publisher, distributor, etc. (see 4E)

If the other data are romanized, give interpolations according to the same romanization.

Give all elements in the other areas (other than the key-title (see 8C) and titles and quotations in notes) in the language and script of the cataloguing agency.

(1.0E)

■APPLICATION   In the mathematical data area use the language and script of the cataloguing agency; an original statement of scale, projection, etc., may be transcribed in a note (7B8).

0F   Inaccuracies   (3.0F)

0F1   Transcribe an inaccuracy or a misspelled word as it appears in the item. Follow such an inaccuracy by ₁*sic*₁ or by the abbreviation *i.e.* and the correction within square brackets. Supply a missing letter or letters in square brackets.   (1.0F)

> Sketth ₁sic₁ of the Hessian attack on Fort Washington

> First map of clasified ₁sic₁ streets and waterways at Port Charlotte

> Cos mos : a spimal ₁i.e. spiral₁ map of universe showing U.S. history and human lifelines

> R.C. Booth Enterp₁r₁ises

0G   Accents and other diacritical marks   (3.0G)

0G1   Add accents and other diacritical marks that are omitted from data found in the source of information in accordance with the usage of the language used in the context.
(1.0G)

0H   Initials, etc.

Record initials, initialisms, and acronyms without internal spaces, regardless of how they are presented in the source of information. Apply this provision also whether or not these elements are presented with full stops.

> Pel battesimo di S.A.R. Ludovico . . .

> KL Ianuarius habet dies xxxi

> Monasterij B.M.V. campililioru₁m₁

> J.J. Rousseau

Treat an abbreviation consisting of more than a single letter as if it were a distinct word, separating it with a space from preceding and succeeding words or initials.

Ph. D.

Ad bibliothecam PP. Franciscan, in Anger

Mr. J.P. Morgan

If two or more distinct initialisms (or sets of initials), acronyms, or abbreviations appear in juxtaposition, separate each from the other with a space.

M. J.P. Rabaut
  (i.e., Monsieur J.P. Rabaut)

par R.F., s. d. C. . . .
  (i.e., par Roland Freart, sieur de Chambray . . .)   (1SBD(A))

0J Items with several chief sources of information   (3.0H)

0J1 Single part items. Describe an item in one physical part from the first occurring chief source of information or that one that is designated as first, unless one of the following applies:

a) Prefer a chief source of information bearing a later date of publication, distribution, etc.
b) If the chief sources present the item in different aspects (e.g., as an individual item and as part of a multipart item), prefer the one that corresponds to the aspect in which the item is to be treated.
c) For items that contain text for which there are chief sources of information in more than one language or script, prefer (in this order):
    i)   the source in the language or script of the text if there is only one such language or script or only one predominant language or script
    ii) the source in the original language or script of the work if the text is in more than one language or script, unless translation is known to be the purpose of the publication, in which case use the source in the language of the translation
    iii) the source in the language or script that occurs first in the following list: English, French, German, Spanish, Latin, any other language using the roman alphabet, Greek, Russian, any other language using the cyrillic alphabet, Hebrew, any other language using the Hebrew alphabet, any other language.

Multipart items. Describe an item in several physical parts from the chief source of information for the first part. If the first part is not available, use the first part that is. If there is no discernible first part, use the part that gives the most information. Failing this, use any part or a container that is a unifying element. Show variations in the chief sources of information of subsequent parts in notes, or by incorporating the data with those derived from the first part.   (1.0H mod.)

0K Description of whole or part

In describing a collection of maps, describe the collection as a whole *or* describe each map (giving the name of the collection as the series), according to the needs of the cataloguing agency. If the collection is catalogued as a whole, but description of the individual parts are considered desirable, see 13. If in doubt about whether to describe the collection as a whole or to describe each part separately, describe the collection as a whole. (3.0J)

1. Description of the collection as a whole

Ordnance Survey of Great Britain one inch to one mile map : seventh series. — Scale 1:63 360. — Chessington ; Southampton : OS, 1952– 74. — maps : col. ; 71 × 64 cm

₁Plans of the Rideau Canal from Kingston Bay to Ottawa / signed by₁ John By . . . ₁et al.₁. — Scales differ. — 1827– 1828. — 28 maps : ms., col. ; 74 × 234 cm or smaller. — Provenance stamps: Board of Ordnance, Inspector General of Fortifications; sheets AA3– 6, 9– 11, 13– 32

Portfolio of Ghana maps. — Scales differ. — Accra : Survey of Ghana, 1961 ₁i.e. 1962₁. — 12 maps : col. ; 36 × 24 cm

2. Description of the collection as a whole—with contents/holdings notes[3] (see also 13C)

Portfolio of Ghana maps. — Scales differ. — Accra : Survey of Ghana, 1961 ₁i.e. 1962₁. — 12 maps : col. ; 36 × 24 cm. — Contents: 1. Ghana administrative — 2. Ghana physical — 3. Ghana vegetation zones — ₁etc.₁

3. Description of one map—separate description (see also 13B)

Banbury / Ordnance Survey. — ₁Ed.₁ B. — Scale 1:63 360. — Southampton : OS, 1968. — 1 map : col. ; 71 × 64 cm. — (Ordnance Survey of Great Britain one inch to one mile map : seventh series ; sheet 145). — ''Fully revised 1965– 66''

Locks and dams at Merrick Mills, sect. no. 5 / ₁signed by₁ John By. — Scale ₁1:1 050₁. — 1827 Oct. 25. — 1 map : ms., col. ; 65 × 75 cm. — (₁Plans of the Rideau Canal from Kingston Bay to Ottawa / signed by₁ John By ; sheet AA29). — Provenance stamps: Board of Ordnance, Inspector General of Fortifications

Ghana vegetation zones / compiled, drawn, and photolithographed by Survey of Ghana. — Scale 1:2 000 000. — Accra : The Survey, ₁1961?₁. — 1 map : col. ; 36 × 24 cm. — (Portfolio of Ghana maps ; 3)

4. Description of one map—multilevel description (see also 13F)

Ordnance Survey of Great Britain one inch to one mile map : seventh series. — Scale 1:63 360. — Chessington ; Southampton : OS, 1952– 1974. — 190 maps : col. ; 71 × 64 cm
  Sheet 145: Banbury. — ₁Ed.₁ B. — 1968. — 1 map. — ''Fully revised 1965– 66''

---

3. The Ordnance Survey and Rideau Canal map examples have not been illustrated here as they are considered too large to be handled with a contents/holdings note.

[Plans of the Rideau Canal from Kingston Bay to Ottawa / signed by]
John By . . . [et. al.]. — Scales differ. — 1827–1828. — 28 maps : ms.,
col. ; 74 × 234 cm or smaller
    Sheet AA29: Locks and dams at Merrick Mills, sect. no. 5. — Scale
[1:1 050]. — 1827. — 1 map ; 65 × 75 cm. — Provenance stamps:
Board of Ordnance, Inspector General of Fortifications

    Portfolio of Ghana maps. — Scales differ. — Accra : Survey of
Ghana, 1961 [i.e. 1962]. — 12 maps : col. ; 36 × 24 cm
    Sheet 3: Ghana vegetation zones / compiled, drawn, and
photolithographed by the Survey of Ghana. — Scale 1:2 000 000. — [1961?].
— 1 map.   (3.0J mod.)

1   TITLE AND STATEMENT OF RESPONSIBILITY AREA

Contents:

1A	Preliminary rule	
1B	Title proper	
1C	General material designation	
1D	Parallel titles	
1E	Other title information	
1F	Statements of responsibility	
1G	Items without a collective title	(3.1)

■APPLICATION   Select the method of description for cartographic items consisting of multiple maps (components; or, primary and ancillary maps), or multiple parts according to rules 0K and 1G. Based on this choice the description of the title area, mathematical data area, and physical description area must be coordinated to ensure consistency in the cataloguing record as a whole. (See Appendix G, example 35.)

1A   Preliminary rule   (3.1A)

1A1   Punctuation. For instruction on the use of spaces before and after prescribed punctuation, see 0C1.
    Precede the title of a supplement or section (see 1B9) by a full stop.
    Enclose the general material designation in square brackets.
    Precede each parallel title by an equals sign.
    Precede each unit of other title information by a colon.
    Precede the first statement of responsibility by a diagonal slash.
    Precede each subsequent statement of responsibility by a semicolon.
    For the punctuation of this area for items without a collective title, see 1G.   (3.1A1)

    Title proper [general material designation] = parallel title = parallel title
/ statement of responsibility

    Title proper [general material designation] : other title information :
other title information / statement of responsibility

    Title proper [general material designation] : other title information =
parallel title : other title information / statement of responsibility

Title proper ₁general material designation₁ / statement of responsibility
= parallel title / statement of responsibility

Title proper ₁general material designation₁ / statement of responsibility ;
second statement of responsibility ; third statement of responsibility

Title proper ₁general material designation₁

1A2   Sources of information. Take information recorded in this area from the chief source of information for the material to which the item being described belongs. Enclose information supplied from any other source in square brackets.

Record the elements of data in the prescribed order, even if this means transposing data, unless case endings are affected, or the grammatical construction of the data would be disturbed, or one element is inseparably linked to another. In the latter cases, transcribe the data as found. (1.1A2)

1B   Title proper (3.1B)

1B1a   Transcribe the title proper exactly as to wording, order and spelling, but not necessarily as to punctuation and capitalization. Give accentuation and other diacritical marks that are present in the chief source of information (see also 0G1). (1.1B1)

Historical north England

A map of the county of Essex

Road map of 50 miles around London

England & Wales

The Edinburgh world atlas, or, Advanced atlas of modern geography

Bouguer gravity anomaly map of Tennessee

Františkovy Lázně orientační plán

British maps of the American Revolution

The Faber atlas

Map of Middle Earth

Geographia marketing and sales maps of Europe

Projected land use maps year 2000, Sulphur Basin

USAF lunar wall mosaic

Bahamas air navigation chart

16-inch sculptural relief globe

1978 stream evaluation map, State of North Dakota

■APPLICATION

"Order" of the title proper. On cartographic items where the title information in the cartouche or title block is arranged decoratively and/or other elements of the description

19

(e.g., author, publisher, cf. 1B2) are interspersed with the title information, deduce the logical sequence and then record the title proper in its semantic order. In cases where various title elements are found in separate locations, record them as instructed in the application to 1B8, Scattered title.

> Atlas of Licking Co., Ohio, combination atlas, and 1875 atlas
> (''and 1875 atlas'' *is separated by extensive author and publication information related to the first atlas only*)

Transcription. In general, do not transcribe letters or symbols used in titles to indicate a trademark, a patent, copyright, etc.

> *On source*
> Encyclopedia of amazing true®© facts
> *Transcribe as*
> Encyclopedia of amazing true facts

> *On source*
> © copyright : how to register your copyright . . .
> *Transcribe as*
> Copyright : how to register your copyright . . .

> *On source*
> A survey of SIMULA™ applications . . .
> *Transcribe as*
> A survey of SIMULA applications . . .

Record such symbols only if they constitute the sole title, or they are an integral part of the title and their exclusion would result in ambiguity or distortion.

Transcribe a dedication that forms an integral part of the title proper. (See also 1B4, 1B14, 7B5.)

Series. When describing a series as a whole, the series number (see Glossary, Series designation) which is descriptive of the whole and is prominently displayed on the chief source of information may be included as part of the title proper.

> North West Europe Army/Air 1:250,000, GSGS 4042 / . . .

> Texas 1:50,000, V782 / . . .

> World 1:5,000,000, series 1106 / . . .

If a series title proper (together with the heading, e.g., main entry) is not unique and, for purposes of collocation and identification, additional elements are required, construct a unique title using the guidelines in 1B7. Record the addition in square brackets.

1B1b An alternative title is part of the title proper (see Glossary). Follow the first part of the title and the word *or* (or equivalent) with commas and capitalize the first word of the alternative title. (1.1B1)

> A modern pilgrim's map of the British Isles, or, more precisely, The Kingdom of Great Britain and Northern Ireland

> The West-India atlas, or, A compendious description of the West-Indies

Sciences du jeu du monde, ou, La carte générale contenante les mondes coeleste, terrestre et civile

Mappe monde, ou, Carte générale du globe terrestre

1B1c   However, if the title proper as given in the chief source of information includes the punctuation marks . . . or [ ], replace them by — and ( ), respectively.

(1.1B1)

Getting around— in Germany
(*Source of information reads:* Getting around . . . in Germany)

■APPLICATION   When replacing . . . in the title proper with —, leave a space after the —, unless the dash is at the beginning.

Getting around— in Germany

not   Getting around—in Germany

but   —and then there were none

When data being transcribed for the bibliographic description includes a colon, a slash, or the equals sign, transcribe these marks if the space can be eliminated on either side. Otherwise a comma or a dash (with the space eliminated on both sides) can be substituted for a colon.

Integrated development study Vieux Port/Bassin Louise, Quebec . . .

New York State transportation/planning map

Land use/land cover maps of Texas

1B1d   If the title proper as given in the chief source of information includes symbols that cannot be reproduced by the typographic facilities available, replace them with a cataloguer's description in square brackets. Make an explanatory note if necessary.

(1.1B1)

Tables of error function and its derivative, [reproductions of equations of the functions]

1B2   If the title proper includes a statement of responsibility or the name of a publisher, distributor, etc., and the statement or name is an integral part of the title proper (i.e., connected by a case ending or other grammatical construction), transcribe it as such.

(1.1B2)

Philips' new practical atlas

Champion map of Greater Knoxville, Tennessee

Dolph's map of Cape Coral, Florida

The Grosset world atlas
(*Published by Hammond Incorporated; distributed by Grosset & Dunlap*)

La route Shell

1B3 If the title proper consists solely of the name of a person or body responsible for the item, give such a name as the title proper. (1.1B3)

>   Melbourne and Metropolitan Board of Works

1B4 Abridge a long title proper only if this can be done without loss of essential information. Never omit the first five words of the title proper (excluding the alternative title). Indicate omissions by the mark of omission. (See also 1B13.) (1.1B4)

■APPLICATION If the dedication forms an integral part of the title proper and precedes the title proper (in part or in whole), do not omit the first five words of the dedication. (See also 1B14.)

>   To the Right Honourable the Lords Commissioners . . .

>   To Sir Watkin Williams Wynn Bart . . .

1B5 If a letter or word appears only once but the design of the chief source of information makes it clear that it is intended to be read more than once, repeat the letter or word without the use of square brackets.

>   *Chief source of information*
>   Canadian BIBLIOGRAPHIES canadiennes

>   *Transcription*
>   Canadian bibliographies = Bibliographies canadiennes

If the first level of description is used (see 0D1), the transcription is: (1.1B5)

>   Canadian bibliographies

1B6 If a title proper includes separate letters or initials without full stops between them, record such letters without spaces between them.

>   UBD detailed street map of Sydney city & suburbs

If such letters or initials have full stops between them, record them with full stops but without any internal spaces. (1.1B6)

>   The U.B.D. map of Wagga Wagga

1B7 If the item lacks a title, and if no title can be found on accompanying material, or a reference source, or elsewhere, devise a brief descriptive title. Always include the name of the area covered in the supplied title. Enclose such a supplied or devised title in square brackets. (1.1B7 mod., 3.1B4 mod.)

>   [Map of Ontario]

>   [Lunar globe]

>   [Gravity anomaly map of Canada]

>   [Carte de la lune]

>   [Braille world atlas]

22

ₗRelief model of California showing vegetationₗ

ₗNautical chart of the coast of Maine from Cape Elizabeth to Monhegan Islandₗ

■APPLICATION   When supplying a title use wherever practicable the language and script of the item in hand. If this is not possible, use the language of the cataloguing agency.

1.   Include the name of the area and main subject (if any) in a constructed title proper for single items. Use terms taken from the item itself, e.g., from the legend or notes. Use natural language order in a supplied title.

ₗGravity anomaly map of Canadaₗ

not   ₗCanada – gravity anomaly mapₗ

ₗMining claims in parts of Gloucester and Restigouche counties, New Brunswickₗ

not   ₗGloucester and Restigouche counties, N.B. – mining claimsₗ

2.   *Series.* If a clearly identifiable and consistently used series title is provided either on the items or in the publishers' literature and indexes, use this title as the title proper. In the latter case enclose it in square brackets.

If there is no readily identifiable series title (e.g., considerable variation in title, or no title at all) construct a unique title using the guidelines below and use this as the title proper for the series.

Guidelines for constructing a unique title for series. When constructing a unique title for a map series, choose only the title elements from the following list which will uniquely identify the series. Record them in the sequence indicated. (See also 0E and 0J.)

If punctuation is required use commas.

Table of sequence of title elements for series

The choice of actual terms (area and subject) used in the constructed unique title should be taken in the following order from the maps themselves (from the legend or notes), from accompanying materials, or from publishers' catalogues, etc.

SEQUENCE	TITLE ELEMENT
1	Area
2	Subject
3	Scale (record as an RF, and only if consistent on all sheets)
4	Series number
5	Corporate body
6	Edition
7	Date (of publication)

1B8a If the title proper appears in two or more languages or scripts, record as the title proper the one in the language or script of the main content of the item. If this criterion is not applicable, choose the title proper by reference to the order of titles on, or the layout of, the chief source of information. Record the other titles as parallel titles (see 1D). (See also 0E and 0J1.) (1.1B8 mod.)

■ APPLICATION

Tête-bêche/back-to-back works. For cartographic works published back to back, in which one map/atlas, etc., is a translation of the other, the work may be treated as one entity, or separate entries may be made for each one. In the former case choose one title as the title proper and record the other as a parallel title. (See 0J, 1D, 1F11.)

1B8b If the chief source of information bears more than one title, in the same language and script, choose the title proper on the basis of the sequence or layout of the titles. If these are insufficient to enable the choice to be made or are ambiguous, choose the most comprehensive title. (See Figure 2.) (3.1B3 mod.)

■ APPLICATION

1. *Choice of title.* If the item has more than one title in the same language and there is doubt as to which should be chosen as the title proper, use the following "Table of title location in order of preference" when taking the steps listed below. The intent is to provide a title that includes an expression of the area covered by, and, if applicable, the subject matter of, the cartographic item.

Table of title location in order of preference

---

a) A title located within the neat line or border of the main map, etc.
b) A title located on the recto of the item outside the neat line or border of the main map, etc.
c) A panel title (recto or verso)
d) A title located on the verso of the item
e) A title located on a cover, container, etc.

Step 1: Consider all of the titles occurring in the locations cited in the table giving precedence to the title which includes the most precise expression of *both* area and subject. If this fails, apply Step 2.

Step 2: Consider all of the titles appearing in the locations cited in the table, giving precedence to the title containing an area element which occurs in the most preferred location (a to e). If this fails, apply Step 3.

Step 3: Select the title in accordance with the order of preference of the table.

For globes and atlases the respective provisions of rules 0B3 and 0B4 have precedence in the selection of the chief source of information and thereby the selection of title.

For multipart cartographic items the provisions of rule 0J have precedence, e.g., a formal, separate title sheet may be given preference as a unifying element, over all other title locations.

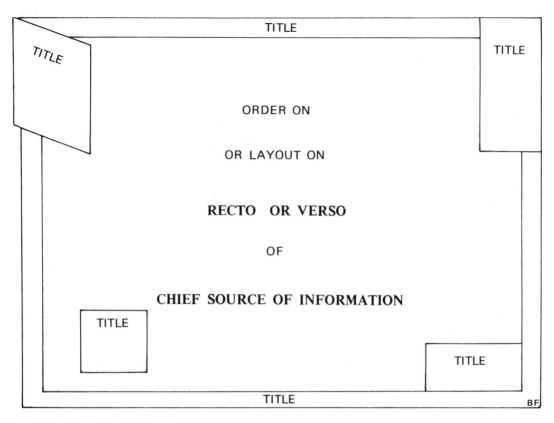

Figure 2   1B8   Choice of title proper when more than one title

2. *Scattered title elements.* If there is no title consisting of area and subject in the chief source of information, but the elements are scattered over the item,[4] construct a title proper in natural reading order, usually based on the following order of preference (if necessary supply connectives in square brackets):

a) any element distinguished by typography and/or position (disregarding any variation in typography which is purely decorative)

b) any element containing the type of format followed by the preposition *of,* e.g., Map of . . .

c) any element composed of, or containing the area

d) any element composed of, or containing the subject

e) any element composed of, or containing the scale in a position warranting or suggesting inclusion in the title

f) any element composed of, or containing the date in a position warranting or suggesting inclusion in the title.

For series, follow the order given in 1B7 Application.
Where punctuation is appropriate use commas.

---

4. See also 1B1, Application regarding titles arranged decoratively.

3.   *Overprint.* If an item has an overprinted title, record it. If the overprinted title does not include any indication of area or subject, follow rule 1E6. Do not include the area in square brackets if it is taken from the title of the base map.

1B9   If the title proper for an item that is supplementary to, or a section of, another item appears in two or more parts not grammatically linked, record the title of the main work first, followed by the title(s) of the supplementary item(s) or section(s) in order of their dependence. Separate the parts of the title proper by full stops.   (1.1B9)

> Carte du cours du fleuve de Saint-Laurent depuis Québec jusqu'à la mer
> en deux feuilles. Ire feuille : depuis Québec jusqu'à Matane

1B10   If the title proper includes a statement of the scale, include it in the transcription.
                                                                                   (3.1B2)
> Topographic 1:500 000 low flying chart
>
> New "half-inch" cycling road maps of England and Wales
>
> Plan-guide de Liberville au 1:10 000
>
> 1:100,000-scale metric topographic map of St. Clair County, Michigan
>
> World 1:5,000,000
>
> A topographical map of Wiltshire, on a scale of 2 inches to a mile
>
> Vegetation map of Collie, latitude 33°00' to 34°00'S, longitude 115°30'
> to 117°00'E, scale 1:250 000
>
> Australia 1:500 000 relief map

■APPLICATION   When the scale is included as part of the title (either title proper or other title information), transcribe it exactly as to punctuation and spacing.

1B11   If the title page of an atlas bears both a collective title and the titles of individual works, give the collective title as the title proper and give the titles of the individual works in a contents note (see 7B18).   (2.1B2 mod.)

Early cartographic items

1B12   If an early printed atlas has no title page and if none of the sources named in 0B5 furnishes a title proper, supply a title according to rule 1B7 Application. As many of the opening words of any prefatory text as required may be recorded in a note if needed for further identification.   (2.14A mod.)

1B13   In abridging a long title proper (see 1B4), omit first any alternative title and the connecting word (*or,* etc.), then omit inessential words or groups of words. Indicate omissions by the mark of omission.   (2.14B)

> *Title appears as*
>   Nouvel atlas des enfants, ou Principes clairs pour apprendre facilement
> et en fort peu de terms la géographie, suivi d'un traite méthodique de la

sphère, qui explique le mouvement des astres, les divers sistèmes du monde, & l'usage des globes; enrichi de xxiv cartes enluminées

Title proper recorded as
>   Nouvel atlas des enfans, . . . suivi d'un traite méthodique de la sphère
>   . . .

1B14   Always omit mottoes, quotations, dedications, statements, etc., appearing on the chief source of information that are separate from the title proper.   (2.14C mod.)

■APPLICATION   Do not use the mark of omission in such cases. Use the note area to record or describe this kind of information if it is considered important.

1B15   Do not treat additions to the title, even if they are linked to it by a preposition, conjunction, prepositional phrase, etc., as part of the title proper.   (2.14D)

Title appears as
>   Atlas of the Chinese Empire containing separate maps of the eighteen
>   provinces of China Proper on the scale of 1:3 000 000 and of the four
>   great dependencies on the scale of 1:7 500 000 together with an index to
>   all the names on the maps and a list of all protestant mission stations, etc.

Title proper recorded as
>   Atlas of the Chinese Empire

■APPLICATION   If such additions to the title constitute a formal statement of the contents of the work, however, record them in a note (7B18).

1B16   Transcription of certain letters. When transcribing information from works published before 1801, do not add accents and other diacritical marks that are not present in the source.

In general transcribe letters as they appear in the text. However, convert earlier forms of letters (e.g., earlier forms of d, r, and s) and earlier forms of diacritical marks to their modern form. Spell out ligatures unless modern usage dictates otherwise. Transcribe β as ss, and transcribe I as I when it is used to indicate "ii" (e.g., as the genitive case in Latin). (For the transcription of i/j and u/v, see below.) When there is any doubt as to the correct conversion of elements to modern form, transcribe them from the source as exactly as possible.

Capitalize or lowercase according to the rules for capitalization in *AACR2* Appendix A. (For roman numerals, see Appendix J.) When the rules for capitalization require changing i/j or u/v from upper to lowercase or vice versa, follow these instructions.

In texts which do not distinguish between i and j or between u and v or w, follow the usage of the text. A text, for example, may consistently give ij as in "filijs"; or, as commonly in early Italian books, initial u may predominate instead of v. If the usage of the text is in doubt, follow these instructions:

Transcribe into lowercase:

I or J as i	U or V as u (but transcribe U or V in initial
II as ii	position as v)
IJ as ij	VV as uu (or vv in initial position)

Transcribe into capitals:

i as I	u or v as V
j as J	uu or vv as VV (i.e., two capital V's)

Treat gothic capitals in the forms J and U as I and V. (In "modern" gothic where lowercase i and j are distinguished, transcribe the gothic capitals according to the lowercase usage.)   (2.14E mod.)

1C *Optional addition.* **General material designation**   (3.1C)

1C1 Choose one of the lists of general material designations given below. The following general material designations are recommended for British and North American use. If general material designations are to be used in cataloguing, British agencies should use terms from list 1 and North American agencies, terms from list 2.

LIST 1	LIST 2[5]
cartographic material	{ map globe
graphic	{ art original chart filmstrip flash card picture slide technical drawing transparency
machine-readable data file	machine-readable data file
manuscript	manuscript
microform	microform
motion picture	motion picture
multimedia	kit
music	music
object	{ diorama game microscope slide model realia
sound recording	sound recording
text	text
videorecording	videorecording

---

5. The following rules apply to list 2: (1) use *map* for cartographic charts, not *chart;* (2) for material treated in *AACR2* chapter 8, use *picture* for any item not subsumed under one of the other terms in list 2; (3) use *technical drawing* for items fitting the definition of this term in *AACR2* Glossary, Appendix D; for architectural renderings, however, use *art original* or *picture,* not *technical drawing;* (4) use *kit* for any item

Use the terms from the list chosen in all descriptions for which general material designations are desired.[6]   (1.1C1)

Central Europe ₍GMD₎

Camden's Britannia, 1695 ₍GMD₎

Decca aeronautical plotting chart ₍GMD₎

☐ POLICIES

British Library
The option is not applied.

Library of Congress
The option is not applied to cartographic materials.

National Library of Australia
The Map Collection does not apply this option. The Cataloguing
Department uses List 2 for any cartographic items it handles, e.g., atlases.

National Library of New Zealand
The option is not applied to cartographic materials.

National Map Collection, PAC
The option is not applied.

1C2 If an item consists of material falling within one category in the list chosen, add the appropriate designation immediately following the title proper.

Town of Preston, with views of principal business buildings
₍microform₎

In the case of items having no collective title, add the appropriate designation immediately following the last title of a group of titles by the same author. If there are several authors, add the designation following the last statement of responsibility appertaining to a group of titles by different authors (see 1G3).   (1.1C2)

1C3 If the item is a reproduction in one material of a work originally presented in another material (e.g., a text as microform; a map on a slide), give the general material designation appropriate to the material being described (e.g., in the case of a map on a slide, give the designation appropriate to the slide).   (1.1C3)

1C4 If an item contains parts belonging to materials falling into two or more categories, and if none of these is the predominant constituent of the item, give either *multimedia* or *kit* as the designation (see 1C1 and 10).   (3.1C2)

---

containing more than one type of material if the relative predominance of components is not easily determinable and for the single-medium packages sometimes called "lab kits."

6. In all subsequent examples, other than those directly illustrating general material designations, the general material designation when indicated is given as ₍GMD₎. The use of ₍GMD₎ in examples does not imply that a designation is required.

Changing Africa ₁multimedia₁

or   Changing Africa ₁kit₁

☐ POLICIES

British Library
The rule is not applied.

Library of Congress
The term *kit* is used.

National Library of Australia
The term *kit* is used.

National Library of New Zealand
The term *kit* is used.

National Map Collection, PAC
The rule is not applied.

1D   Parallel titles   (3.1D)
See also 1B8 Application, Tête-bêche.

1D1   Record parallel titles in the order indicated by their sequence on, or by the layout of, the chief source of information.   (1.1D1)

Suomi ₁GMD₁ = Finland

International map of natural gas fields in Europe ₁GMD₁ = Carte internationale des champs de gaz naturel en Europe

Mobil street map of Durban ₁GMD₁ = Straatkaart van Durban

1D2   In preparing a second-level description (see 0D2), give the first parallel title. Give any subsequent parallel title that is in English. If no title is in English and the title proper is in a nonroman script, give the parallel title that is (in order of preference) in French, German, Spanish, Latin, or any other roman alphabet language.

Business atlas of Western Europe ₁GMD₁ = Atlas des affaires de l'Europe occidentale

Fennia : suuri Suomi-kartasto ₁GMD₁ = Kartverk over Finland = Finland in maps

Israel in maps for the blind ₁GMD₁ = ₁Yiśrael be-mapot la-'iyrim₁
(Transliterated parallel title)

Strassenkarte der Schweiz ₁GMD₁ = Carte routière de la Suisse = Road map of Switzerland

In preparing a third-level description (see 0D3), transcribe all parallel titles appearing in the chief source of information according to the instructions in 1B.   (1.1D2)

Business atlas of Western Europe ₁GMD₁ = Atlas des affaires de
l'Europe occidentale = Handelsatlas Westeuropas = Atlas comercial de
Europa occidental

Strassenkarte der Schweiz ₁GMD₁ = Carte routière de la Suisse = Road
map of Switzerland = Carte stradale della Svizzera

□ POLICIES

British Library
Third-level description is used where practicable.

Library of Congress
Third-level description is used only for materials issued in the United
States.

National Library of Australia
Third-level description is used for all materials where practicable.

National Library of New Zealand
Third-level description is used.

National Map Collection, PAC
Third-level description is used only for Canadian materials but not
necessarily all Canadian materials.

1D3   Record an original title in a language different from that of the title proper
appearing in the chief source of information as a parallel title if the item contains all or
some of the text in the original language, or if the original title appears before the title
proper in the chief source of information. Record an original title in the same language
as the title proper, as other title information (see 1E). In all other cases give the original
title in a note.   (1.1D3)

Sistema Panamericano de carreteras, por países ₁GMD₁ = Pan American
highway system ₁by countries₁
(Contains parallel English legend and Spanish translation)

1D4   Record parallel titles appearing outside the chief source of information in a note
(see 7B5).   (1.1D4)

1E   Other title information   (3.1E)

1E1   Transcribe all other title information appearing in the chief source of information
according to the instructions in 1B.   (1.1E1)

Canada ₁GMD₁ : a pictorial & historical map

Ethelreda's Isle ₁GMD₁ : a pictorial map of the Isle of Ely to
commemorate the 1300th anniversary of the founding of Ely's conventual
church

Motor road map of south-east England ₍GMD₎ : showing truck and other classified roads

Kaunispää-Kopsusjärvi ₍GMD₎ : ulkoilukartta

Peru-Chile, boundary dispute maps, 1544–1879 ₍GMD₎ : photographs of 300 maps

Nicholson's London map ₍GMD₎ : big area and big scale

Upper Chesapeake Bay, satellite image map ₍GMD₎ : NASA LANDSAT imagery 1:500,000, N3849 W7649

Michigan official transportation map ₍GMD₎ : Great Lake State

Map of Ft. Myers, Florida ₍GMD₎ : including detailed insets of Cape Coral, Lehigh Acres, and North Ft. Myers

1E2   Record other title information in the order indicated by the sequence on, or the layout of, the chief source of information.   (1.1E2)

Distribution of the principal kinds of soil ₍GMD₎ : orders, suborders, and great groups : National Soil Survey classification of 1967

1E3   Lengthy other title information. If the other title information is lengthy, either give it in a note (see 7B5) or abridge it.

Abridge other title information only if this can be done without loss of essential information. Never omit the first five words of the other title information. Indicate omissions by the mark of omission.   (1.1E3)

Champion street atlas, Akron, Ohio ₍GMD₎ : including Baberton, Cuyahoga Falls, Fairlawn, Lakemore, Mogadore, Munroe Falls, Norton Stow, Tallmadge, all of Summit County

Environmental geologic atlas of the Texas Coastal Zone-Bay City-Freeport area ₍GMD₎ : environmental geology, physical properties, environments and biologic assemblages, current land use, mineral and energy resources, active processes, man-made features and water systems, rainfall, stream discharge, and surface salinity, topography and bathymetry

The comprehensive atlas & geography of the world ₍GMD₎ : comprising an extensive series of maps, a description, physical and political, of all the countries of the earth : a pronouncing vocabulary of geographical names, and a copious index of geographical positions : also numerous illustrations printed in the text, and a series of coloured engravings representing the principal races of mankind
(Minor title changes distinguish editions of some works, especially before the 20th century)

1E4   If the other title information includes a statement of responsibility or the names of a publisher, distributor, etc., and the statement or name is an integral part of the other title information, transcribe it as such.   (1.1E4)

The complete atlas of modern, classical and celestial maps, together
with plans of the principal cities of the world ₍GMD₎ : constructed and
engraved on steel, under the superintendence of the Society for the
Diffusion of Useful Knowledge, and including all the recent geographical
discoveries, compiled from the latest and most recent sources :
accompanied by alphabetical indexes to the modern and classical maps

Time on our side? ₍GMD₎ : a survey of archaeological needs in Greater
London : report of a joint working party of the Department of the
Environment, the Greater London Council, and the Museum of London

1E5   Transcribe other title information following the title proper or parallel title to
which it pertains.

Guide général de Paris ₍GMD₎ : répertoire des rues, avec indication de
la plus proche station du Métro = General guide to Paris : with repertory
of streets and indication of the nearest Metro station

Périgord du temps passé ₍GMD₎ : 8 estampes rares = 8 rare prints

If there are no parallel titles and if other title information appears in more than one
language or script, give the other title information that is in the language or script of the
title proper. If this criterion does not apply, give the other title information that appears
first. *Optionally,* add the other title information in other languages.   (1.1E5)

☐ POLICIES

British Library
    The option is applied.

Library of Congress
    The option is generally applied.

National Library of Australia
    The option is applied.

National Library of New Zealand
    The option is applied.

National Map Collection, PAC
    The option is applied.

1E6   If neither the title proper nor the other title information includes an indication of
the geographic area covered by the item, or if there is no other title information, add, as
other title information, a word or brief phrase indicating the area covered.

(3.1E2)

Vegetation ₍GMD₎ : ₍in Botswana₎

If the title proper needs explanation, make a brief addition as other title information,
in the language of the title proper.   (1.1E6)

Geology of Texas ₍GMD₎ : ₍index map₎

>Streetmap of Flagstaff, Arizona [GMD] : [walking tours]

>Conference on Industrial Development in the Arab Countries [GMD] : [proceedings]

■APPLICATION If the title proper, other title information, or variant title does not indicate the subject content of the item, it may be added as other title information in the language of the title proper. If this is not practicable (see 0E), it should be in the language of the cataloguing agency.

1F Statements of responsibility[7] (3.1F)

1F1 Record statements of responsibility appearing prominently in the item in the form in which they appear there. If a statement of responsibility is taken from a source other than the chief source of information, enclose it in square brackets. (1.1F1)

>Football history map of England and Wales [GMD] : showing . . . the colours and locations of all the clubs in the Football League . . . / compiled by John Carvosso

>The English pilot, the fifth book [GMD] / J. Seller & C. Price

>A map book of West Germany [GMD] / A.J.B. Tussler, A.J.L. Alden

>Road atlas Europe [GMD] / Bartholomew

>World atlas [GMD] / compiled by Rand McNally International

>Ecoregions of the United States [GMD] / by Robert G. Bailey ; prepared in cooperation with the U.S. Fish and Wildlife Service

>Main road map of Cuyahoga County [GMD] / compiled, published, and copyrighted by Commercial Survey Co.

>Philips' new practical atlas [GMD] / director of cartography, Harold Fullard, cartographic editor, B.M. Willett

>The atlas of the earth [GMD] / [editor, Tony Loftas] ; with a foreword by Sir Julian Huxley

>Urban atlas, tract data for standard metropolitan statistical areas [GMD] : Minneapolis-St. Paul, Minnesota / U.S. Department of Commerce, Bureau of the Census [and] U.S. Department of Labor, Manpower Administration

>Oil and natural gas map of Asia [GMD] / prepared under the sponsorship of the United Nations Economic and Social Commission for Asia and the Pacific (ESCAP) ; co-ordinator, V.V. Sastri, with guidance from B.S. Negri ; compilers, L.L. Bhandari . . . [et al.]

---

7. The statement of responsibility relates to persons or corporate bodies who have contributed to the intellectual or artistic content of the cartographic item (e.g., author, cartographer, compiler, engraver, governmental mapping agency, illuminator, reviser, scientific editor, etc.)

Lancaster Township, official zoning map ₍GMD₎ / prepared by the Lancaster County Commission for the Lancaster Township Planning Commission

Route planning, Great Britain ₍GMD₎ / ₍general editor, Roger Edwards ; illustration/typography, Ronald Maddox ; cartography, Map Productions₎

The Wills south coast yachting guide ₍GMD₎ / edited by the Daily express

Flugbild Schweiz ₍GMD₎ = Vue aérienne Suisse = Air view Switzerland / Swissair Photo + Vermessungen AG

Maps & plans of the operations, movements, battles & sieges of the British Army, during the campaigns in Spain, Portugal, and the south of France from 1808 to 1814 ₍GMD₎ / compiled by Lieut. Godwin ; engraved by Jas. Wyld

Glacial map of Tasmania ₍GMD₎ / compiled by E. Derbyshire . . . ₍et al.₎

₍Pocket terrestrial globe₎ ₍GMD₎ / J. Moxon

Mondmapo ₍GMD₎ / laŭ la decidoj de Internacia Komisiono por Ordigo de Geografiaj Nomoj ; redaktis, Tibor Sekelj

Wheaton's atlas of British and world history ₍GMD₎ / by the late T.A. Rennard ; editors, H.E.L. Mellersh and B.S. Trinder ; maps prepared by David A. Hoxley

Map catalogue ₍GMD₎ / Ordnance Survey

■APPLICATION   Refer to rule 0.2 for a definition of ''prominence.''
For maps, charts, etc., the whole item is the chief source of information; therefore, if a statement of responsibility appears anywhere on the item, record it.
The following terms, or any combination thereof, are indicative of responsibility (see also 1F12 Application):

> artwork
> by
> cartographer
> cartography by
> compiled *or* recompiled
> corrected
> created *or* recreated
> dedicated by . . . to; *or,* dedicated to . . . by (*not*
>   dedicated to . . .)
> delineated
> designed
> done *or* redone
> drafted *or* redrafted (*if it appears alone on the cartographic*
>   *item*)
> drawn

> edited *or* re-edited
> engraved *(if it appears alone on the cartographic item)*
> made *or* remade
> made up
> prepared
> produced
> revised
> reworked
> surveyed
> updated

1F2   If no statement of responsibility appears prominently in the item, neither construct one nor extract one from the content of the item. (But see also 1F16 and 1F17.)

Do not include statements of responsibility that do not appear prominently in the item in the title and statement of responsibility area. If such a statement is necessary, give it in a note.   (1.1F2)

■APPLICATION   If it is known that the publisher is responsible for the work,[8] but there is no statement of responsibility on the item, a note (7B6) may be made to indicate that the publisher is also responsible for its preparation.

1F3   If a statement of responsibility precedes the title proper in the chief source of information, transpose it to its required position unless case endings would be affected by the transposition. In the latter instance, see 1B2.   (1.1F3)

■APPLICATION   The purpose of this rule is to continue a long-standing practice of transposing clear statements of responsibility from a position at head of title to a position following the title. Occasionally, however, a phrase or a name that is clearly not a statement of responsibility will be found at head of title. Use an ''at head of title'' note for these and also for indeterminate cases (see 7B6).

1F4   Record a single statement of responsibility as such whether the two or more persons or corporate bodies named in it perform the same function or different functions.
                                                                                          (1.1F4)

> Atlas de la lucha de liberación nacional, Nicaragua libre [GMD] / el esfuerzo colectivo del Instituto Geográfico Nacional (I.G.N.) y del Centro del Investigaciones Geográficas

> Región del Biobío, Chile [GMD] / [ha sido elaborado por encargo de la Intendencia de la Región del Bío Bío, en forma conjunta por el Centro Interdisciplinario de Estudios Regionales (CIER) de la Unversidad Católica de Chile, sede regional Talcahuano y la Secretaría Regional de Planificación y Coordinación de la VIII Región (SERPLAC)]

---

8. This knowledge depends on the cataloguer's familiarity with the field and with the preparation and publishing practices of the body concerned. If in doubt, do not make a note.

Atlas des départements français d'outre-mer ₍GMD₎ / réalisé par le
Centre de géographie tropicale du C.N.R.S. Bordeaux-Talence ; avec le
concours des départements de géographie des universités d'Aix-Marseille
II, de Bordeaux III, des centres universitaires des Antilles-Guyane et de la
Réunion, de l'ORSTOM pour l'atlas de la Guyane

Geologic map atlas and summary of economic mineral resources of
Converse County, Wyoming ₍GMD₎ / by Donald W. Lane in collaboration
with Forrest K. Root and Gary B. Glass

New Zealand land resource inventory worksheet ₍GMD₎ / produced for
the National Water & Soil Conservation Organisation by the Water & Soil
Division, Ministry of Works & Development

Atlas esquemático de ciencias geográficas ₍GMD₎ / Cayetano di Leoni
₍y₎ Omar I. Genovese

Census metropolitan area Calgary, Alberta ₍GMD₎ : average household
income for 1971 / the Financial post survey of markets computer income
maps produced in association with Lanpar Limited

■APPLICATION   If the statement of responsibility is in a single phrase, record it as
given on the item. If the statement of responsibility is not in a single phrase, see 1F6.

1F5   When a single statement of responsibility names more than three persons or
corporate bodies performing the same function, or with the same degree of responsibility, omit all but the first of each group of such persons or bodies. Indicate the omission
by the mark of omission (. . .) and add *et al.* (or its equivalent in the nonroman scripts)
in square brackets.   (1.1F5)

Presettlement vegetation of Kalamazoo County, Michigan ₍GMD₎ / by
Thomas W. Hodler . . . ₍et al.₎

₍Administrações regionais, atlas₎ ₍GMD₎ / Prefeitura Municipal de São
Paulo, COGEP-Coordenadoria Geral de Planejamento . . . ₍et al.₎

■APPLICATION   If the cartographic design results in a diffuse arrangement of information, it may be inappropriate to record the first named person in a single statement of
responsibility reading from left to right. If one of the names is distinguished by placement relative to the others or by typography, record that name and omit the remainder.
Make a note if considered necessary.

1F6   If there is more than one statement of responsibility, record them in the order
indicated by their sequence on, or by the layout of, the chief source of information. If
the sequence and layout are ambiguous or insufficient to determine the order, record the
statements in the order that makes the most sense. If statements of responsibility appear
in sources other than the chief source, record them in the order that makes the most
sense.   (1.1F6)

The national trucker's scale and inspection stations atlas ₍GMD₎ / art
and layout by Darlene Moore ; edited and compiled by Faye Cline, Gloria
Watson, Lilli Haywood

GATE international meteorological radar atlas [GMD] / Richard Arkell, Michael Hudlow, Center for Experiment Design and Data Analysis, Washington, D.C. ; other principal contributors G. Austin . . . [et al.]

Climatic atlas of the outer continental shelf waters and coastal regions of Alaska [GMD] / NCC, William A. Brower, Jr., Henry F. Diaz, Anton S. Prechtel ; AEIDC, Harold W. Searby, James L. Wise ; Arctic Environmental Information and Data Center, University of Alaska, Anchorage, Alaska, [and] National Climatic Center, Environmental Data Service, Asheville, North Carolina, [and] National Oceanic and Atmospheric Administration

Rhône-Alpes [GMD] : le portrait d'une région, itinéraires pour une découverte : 140 cartes au 1/100 000 / mises au point pour cet ouvrage par l'Institut géographique national ; préface de Pierre Doueil ; présentation de Roger Frison-Roche et Jacques Soustelle ; textes d'André Lugagne

The Anchor atlas of world history [GMD] : from the Stone Age to the eve of the French Revolution / Hermann Kinder and Werner Hilgemann ; translated by Ernest A. Menze with maps designed by Harald and Ruth Bukor

■APPLICATION   If there are four or more statements of responsibility on an item, it is necessary to record only those persons or bodies judged by their function to have made the most significant contribution. If this is not possible to determine, only the first three on the item, or those whose names are given prominence by typography need be recorded.

1F7   Include titles and abbreviations of titles of nobility, address, honour, and distinction, initials of societies, qualifications, etc., with the names of persons in statements of responsibility if:

a)   such a title is necessary grammatically

   . . . prologo del Excmo. Sr. D. Manuel Frage Iribarne

b)   the omission would leave only the person's given name or surname

   . . . / by Miss Jane

   . . . / by Dr. Johnson

c)   the title is necessary to identify the person

   . . . / by Mrs. Charles H. Gibson

d)   the title is a title of nobility or is a British title of honour (Sir, Dame, Lord, or Lady).

Omit all other titles, etc., from the names of persons in statements of responsibility. Do not use the mark of omission.

   . . . / by Harry Smith
   (*Source of information reads:* by Dr. Harry Smith)   (1.1F7)

■APPLICATION   For early cartographic items include titles and abbreviations of titles of nobility, address, honour, and distinction that appear with names in statements of responsibility.

1F8   Add a word or short phrase to the statement of responsibility if the relationship between the title of the item and the person(s) or body (bodies) named in the statement is not clear.   (3.1F2)

> Maps of the Mid-west ₁GMD₁ / ₁edited by₁ D.M. Bagley

> Official visitor's map and guide to Portland ₁GMD₁ / ₁cartography by₁ Richard J. Williamson ; text by Arthur K. Johnson

> Master plan, city of Grandville, Michigan ₁GMD₁ / prepared by Scott Bagby and Associates ; ₁prepared for₁ Grandville Planning Commission

but

> Northern Ireland ₁GMD₁ : a census atlas / Paul A. Compton in association with . . . ₁others₁

> Atlas zuidoost Utrecht ₁GMD₁ / Atlasprojektgroep

1F9   Replace symbols or other matter that cannot be reproduced by the typographic facilities available with the cataloguer's description in square brackets. Make an explanatory note if necessary.   (1.1F9)

1F10   If an item has parallel titles but a statement or statements of responsibility in only one language or script, give the statement of responsibility after all the parallel titles or other title information.   (1.1F10)

> Taṣmīm ḥaḍarī Īfrān ₁GMD₁ = Plan urbain, Ifrane / publié en 1977 par la Division de la carte
> *(Title in Arabic script, transliterated for catalogue record; remainder of map in French)*

1F11   If an item has parallel titles and a statement or statements of responsibility in more than one language or script, give each statement after the title proper, parallel title, or other title information to which it relates. (See also 1B8a Application, Tête-bêche.)

> Carte géologique du Népal (ouest du Népal) ₁GMD₁ / par J.-M. Remy, 1973 = Geological map of Nepal (west of Nepal) / by J.-M. Remy, 1973

If it is not practicable to give the statements of responsibility after the titles to which they relate, give the statement of responsibility in the language or script of the title proper and omit the others.   (1.1F11)

1F12   Treat a noun phrase occurring in conjunction with a statement of responsibility as other title information if it is indicative of the nature of the work.

If the noun or noun phrase is indicative of the role of the person(s) or body (bodies) named in the statement of responsibility rather than of the nature of the work, treat it as part of the statement of responsibility.

Atlas of Licking Co., Ohio ₍GMD₎ : from actual surveys / by and under the direction of F.W. Beers

In case of doubt, treat the noun or noun phrase as part of the statement of responsibility. (1.1F12)

■APPLICATION   Examples of noun phrases commonly occurring on cartographic items are: drawing by, cartography by, engraving by, geology by, hills by, legend by, outline by, etc.

1F13   When a name associated with responsibility for the item is transcribed as part of the title proper (see 1B2) or other title information (see 1E4), do not make any further statement relating to that name unless such a statement is required for clarity, or unless a separate statement of responsibility including or consisting of that name appears in the chief source of information. (1.1F13)

Foster's cyclists' road map of eastern Ontario ₍GMD₎. —

Gregory's Sydney street directory ₍GMD₎. —

but

Meyers neuer Handatlas ₍GMD₎ / herausgegeben vom Geographisch-Kartographischen Institut Meyer unter Leitung von Adolf Hanle

Diercke Weltatlas ₍GMD₎ / Begründt von C. Diercke
(Name of author appears separately in the chief source of information as well as in the title proper)

1F14   Transcribe a statement of responsibility even if no person or body is named in that statement. (1.1F14)

Map reading and panorama sketching ₍GMD₎ / by an instructor

1F15   Omit statements found in the chief source of information that neither constitute other title information nor form part of statements of responsibility. A phrase such as "with a spoken commentary by the artist" is a statement of responsibility. Statements of responsibility may include words or phrases which are neither names nor linking words (e.g., . . . / written by Jobe Hill in 1812). (1.1F15)

Manuscripts

1F16   *Optional addition.* If the name appended to, or the signature on, a manuscript is incomplete, add to it the name of the person concerned. (4.1F2)

₍Letter₎ 1929 Feb. 8, New York ₍to₎ F. Scott Fitzgerald, Wilmington, Del. ₍GMD₎ / Zelda ₍Fitzgerald₎

₍Letter₎ 1898 July 19, Dorking, Surrey ₍to₎ H.G. Wells, Worcester Park, Surrey ₍GMD₎ / G.G. ₍George Gissing₎

☐ POLICIES
> British Library
> The option is applied.

> Library of Congress
> The option is applied.

> National Library of Australia
> The option is applied.

> National Library of New Zealand
> The option is applied.

> National Map Collection, PAC
> The option is applied.

1F17 If a manuscript lacks a signature or statement of responsibility, supply the name(s) of the person(s) responsible for it, if known. (4.1F3)

> Plan of the attack on Fort William Henry and Ticonderoga ₁GMD₁ : showing the road from Fort Edward, Montcalm's camp and wharf of landing &c. / ₁James Montrésor₁

1G Items without a collective title (3.1G)
For map series see Appendices D and E.

1G1 If a cartographic item lacks a collective title, *either* describe the item as a unit (see 1G2, 1G3, 1G4, and 1G5), *or* make a separate description for each separately titled part (see 1G6), *or* (in certain circumstances) supply a collective title (see 1G3 Application and 1G7). (3.1G1)

1G2 If, in an item lacking a collective title, one work is the predominant part of the item, treat the title of that part as the title proper and name the other parts in a note (see 7B18). (1.1G1)

■ APPLICATION
1. Apply this rule to cartographic items in parts, as appropriate.
2. Apply this rule to single sheet cartographic items containing multiple maps in which one of the maps is clearly predominant, i.e., identified or selected as the main or primary map, and the other maps are supplementary or ancillary to the predominant map.
3. If the component maps are on both sides of the sheet choose one side as the recto and record only those elements concerning it in the description. Record the other map in the "on verso" note (7B18).

Determine "predominance" on the basis of "publisher's preference" as indicated by relative size and scale of parts; placement or prominence of title information; arrangement of constituent parts, ancillary maps, etc.; or the relationship between primary and supplementary or ancillary maps.

1G3   If, in an item lacking a collective title, no one part predominates, record the titles of the individually titled parts in the order in which they are named in the chief source of information, or in the order in which they appear in the item if there is no single chief source of information. Separate the titles of the parts by semicolons if the parts are all by the same person(s) or body (bodies), even if the titles are linked by a connecting word or phrase. If the individual parts are by different persons or bodies, or in case of doubt, follow the title of each part by its parallel titles, other title information, and statements of responsibility and a full stop followed by two spaces.   (1.1G2)

> Illustrated historical atlas of the county of Simcoe, Ont. ₁GMD₁ / H.
> Belden & Co., Toronto, 1881. And Hogg's map of the county of Simcoe /
> compiled and published by John Hogg, Collingwood, Ont., 1971. Simcoe
> farmer's directory, Union Pub. Co., 1890
> *(Transcribed from title page in order of presentation)*

> Grand Teton ; Yellowstone National Park ₁GMD₁
> *(Both maps produced by the same body)*

> Daily mail motor road map of London and twelve miles round. Motor
> road map of south-east England ₁GMD₁
> *(Maps produced by different bodies)*

> Dissegno della fabrica fatta ad uso delle fiere di Verona nell'anno 1722
> Pierantonio Berno forma in Verona. Pianta della fiera di Verona ₁GMD₁

■APPLICATION   This rule also applies to maps in components as well as multipart items. (See Glossary, Component.)

If the cartographic item consists of more than three components, or equal parts, supply a collective title (see 1B7) and record the individually titled components or parts in a note (see 7B18 Application).

If the cartographic item consists of two or three components, or equal parts, record the individually titled components or parts in an order based on the following factors:

1) the arrangement of components on the sheet
2) institutional policy
3) the presence or absence of indexing information
4) publisher's preference as indicated by the order in which the parts or components are named on the chief source of information
5) the reference value or uniqueness of one of the components or parts, e.g., the choice or selection of a thematic map over a general map
6) the size and scale of components or parts.

1G4   Make the relationship between statements of responsibility and the parts of an item lacking a collective title and described as a unit clear by additions as instructed in 1F8.   (3.1G3)

1G5   If, in an item lacking a collective title, more than one (but not all) of the separately titled parts predominate, treat the predominating parts as instructed in 1G3, and name the other parts in a note (see 7B18).   (1.1G4)

42

1G6 If desired, make a separate description for each separately titled part of an item lacking a collective title. Link the separate descriptions with a note (see 7B21).

(3.1G4)

1G7 If a cartographic item lacking a collective title consists of a large number of physically separate parts, supply a collective title as instructed in 1B7.

(3.1G5)

ₗMaps of Denmarkₗ

ₗCollection of tourist maps of Thailandₗ

2   EDITION AREA

Contents:

2A   Preliminary rule   (3.2A)

2A1   Punctuation. For instructions on the use of spaces before and after prescribed punctuation, see 0C1.

Precede this area by a full stop, space, dash, space.

Precede a subsequent edition statement by a comma.

Precede the first statement of responsibility following an edition or subsequent edition statement by a diagonal slash.

Precede each subsequent statement of responsibility by a semicolon.   (3.2A1)

. — Edition statement

. — Edition statement, second edition statement

. — Edition statement / statement of responsibility

. — Edition statement / statement of responsibility, second edition statement / statement of responsibility

— Edition statement / statement of responsibility ; second statement of responsibility ; third statement of responsibility

2A2   Sources of information. Record in this area information taken from the chief source of information or from any other source specified for this area in the following rules. Enclose any information supplied from any other source in square brackets.

(1.2A2)

2B   Edition statement   (3.2B)

See also 2B8 and 2B9 for early cartographic items.

2B1　Transcribe a statement relating to an edition of a work that contains differences from other editions, or that is a named revision of that work as instructed in the following rules.　　　　(3.2B1)

éd. 1	9th ed. ₍cover₎, 12th ed. ₍maps₎
Éd. 1	18ª ed. ₍cubierta₎, 6ª ed. ₍mapas₎
1ᵉʳ ed.	Rev. 1976
1ª ed. ampliada	As rev. by by-law no. 19, of 1963
1st Tuttle ed.	1974 new ed.
2ᵉ éd.	1974/75 ed.
2nd ed.	15th revision
3rd U.S. ed.	1st ed. repr. with revisions to date, 1967
3rd impression rev.	4th ed., 1978 ed.
54. Aufl.	8th ed. 1980
Ed. 5-SK	22nd ed. Feb. 1980
Advance ed.	Repr. March 1933 with corrections
Amended 5/31/76	Photorevised 1978
Comprehensive ed.	Minor corrections made 1974
Edizone speciale Shell Italiana	Provisional issue
Ed. latino-americana	Preliminary ed.
Facsim. ed.	New perspective ed. incl. zip codes
Rev. et corr.	261. Aufl., Ausg. mit Karte zur
Rev. and enl.	Heimatskunde

Transcribe the edition statement as found on the item. Use standard abbreviations (see Appendix H) and numerals in place of words (see Appendix J).　　　　(1.2B1)

On source of information
Ny udgave
Transcription
Ny udg.

On source of information
Second edition
Transcription
2nd ed.

Special ed. of the Atlas of Darke County, Ohio 1857

1ª ed., en el sis-cents aniversari de la seva realització, 1375–1975

Rev. in 1979 from Dept. of Transportation 1:24,000 scale quadrangle maps, highway construction plans, municipal boundary maps, and various other sources

■APPLICATION   When a scale is included as part of the edition statement, transcribe it exactly as to punctuation and spacing.

2B2   If the edition statement consists solely or chiefly of characters that are neither numeric nor alphabetic, record the statement in words in the language and script of the title proper and enclose them in square brackets.

⌐Three asterisks⌐ ed.

If the edition statement consists of a letter or letters and/or a number or numbers without accompanying words, add an appropriate word or abbreviation.

(1.2B2)

3e ⌐éd.⌐

⌐Ed.⌐ B

2B3   In case of doubt about whether a statement is an edition statement, take the presence of such words as *edition, issue, version* (or their equivalents in other languages) as evidence that such a statement is an edition statement, and record it as such.

(1.2B3)

South-west gazette ⌐GMD⌐. — Somerset ed.

Subbuteo table soccer ⌐GMD⌐. — World Cup ed.

■APPLICATION   Terms, phrases, or codes which provide information on the bibliographical status of the item, but which are not recorded in the edition area may be recorded in a note.

2B4   *Optional addition.* If an item lacks an edition statement but is known to contain significant changes from previous editions, supply a suitable brief statement in the language and script of the title proper and enclose it in square brackets.

(3.2B3, 1.2B4)

⌐5th ed.⌐

⌐Nouv. éd.⌐

⌐New ed.⌐

⌐3e éd.⌐

⌐2nd ed., partly rev.⌐

☐POLICIES

British Library
   The option is not applied. Any edition information that is supplied is given in the note area.

Library of Congress
The option is not applied to any case of merely supposed differences in issues that might make them different editions. The option is applied to make a distinction among those cases where differences are manifest, but their catalogue records would show exactly the same information from the title and statement of responsibility area to the series area inclusive.

National Library of Australia
The option is not applied. Any edition information supplied is given in the note area.

National Library of New Zealand
ₗRev. ed.ₗ or ₗNew ed.ₗ are supplied; numerical statements are not supplied. Edition statements are always accompanied by an edition note.

National Map Collection, PAC
The option is not applied. Any edition information that is supplied is given in the note area.

2B5   If an edition statement appears in more than one language or script, record the statement that is in the language or script of the title proper. If this criterion does not apply, record the statement that appears first (but see also 2B6).   (3.2B4, 1.2B5)

Carte géologique internationale de l'Europe ₗGMDₗ = International geological map of Europe. — 3ᵉ éd.

■APPLICATION   *Optionally,* give the parallel statements, each preceded by an equals sign.

□POLICIES
National Map Collection, PAC
Give parallel edition statements if they are in English and/or French.

2B6   Serials. If an edition statement appears in two or more languages or scripts, give the statement that is in the language or script of the title proper, *or,* if this criterion does not apply, the statement appearing first, and *optionally* the parallel statement(s), each preceded by an equals sign.   (12.2B3 mod.)

Canadian ed. = Ed. canadienne

□POLICIES
British Library
The option is applied.

Library of Congress
The option is applied.

National Library of Australia
The option is applied.

National Library of New Zealand
The option is applied.

National Map Collection, PAC
The option is applied.

2B7   If an item lacking a collective title and described as a unit contains one or more parts with an associated edition statement, record such statements following the titles and statements of responsibility to which they relate, separated from them by a full stop.

(3.2B5)

> Le western / textes rassemblés et présentés par Henri Agel. Nouv. éd.
> Évolution et renouveau du western (1967–1968) / par Jean A. Gili ₍GMD₎

2B8   If the edition statement is an integral part of the title proper, other title information, or statement of responsibility, or if it is grammatically linked to any of these, record it as such and do not make a further edition statement.   (2.15B)

> The small English atlas, being a new and accurate sett ₍sic₎ of maps of all the counties in England and Wales ₍GMD₎ / Thomas Jefferys and T. Kitchin

> Atlas abrégé et portatif . . . ₍GMD₎ / par M. Labbé de Gourné ; revû, corrigé et augmenté sur les nouvelles observationes astronomiques faites en 1741 par M M. Tchirikcow et De l'Isle

> A general atlas describing the whole universe ₍GMD₎ : being a complete and new collection of the most approved maps extant : corrected with the utmost care, and augmented from the latest discoveries, down to 1782 : the whole being an improvement of the maps of d'Anville and Robert ₍de Vaugondy₎ . . .

> World game edition dymaxion sky-ocean map ₍GMD₎ : the Fuller projection / Buckminster Fuller and Shoji Sadao, cartographers Loudoun edition, northern Virginia community shelter plan

but

> Atlas minimus, or, A new set of pocket maps of the several empires, kingdoms and states of the known world, with historical extracts relative to each ₍GMD₎ / drawn and engrav'd by J. Gibson from the best authorities ; revis'd, corrected and improv'd, by Eman: Bowen. — A new ed., corr.

> The third centenary edition of Johan Blaeu Le grand atlas, ou, Cosmographie blaviane ₍GMD₎. — Facsim. ed. in 12 v.
> (*On t.p.:* Facsimile edition in twelve volumes)

> Gregory's metric edition, Australia ₍GMD₎ / production, Clive Barras.
> — 8th ed.

2B9   Early cartographic items. In general, for early cartographic items, record an edition statement as it is found in the item. If an exact transcription is not desired, use standard abbreviations and arabic numerals in place of words as instructed in 2B.

(2.15A mod.)

Nunc primum in lucem aedita

Editio secunda auctior et correctior

Cinquième édition

or 5e éd.

2C Statements of responsibility relating to the edition (3.2C)

2C1 Record a statement of responsibility relating to one or more editions, but not to all editions, of a cartographic item following the edition statement if there is one. Follow the instructions in 1F for the transcription and punctuation of such statements.

(3.2C1 mod., 1.2C1 mod.)

3rd ed. / with maps redrawn by N. Manley

Rev. 1977 / J.B. Phillips

10th ed. / edited by Dana and Helen Dalrymple

Ausg. Baden-Württemberg, 1. Aufl. / [unter Mitwirkung von Friedrich Pfrommer und schulgeographischer Fachkreise bearbeitet von der Verlagsredaktion]

2nd ed. / prepared by the Cartographic Dept. of the Clarendon Press

2C2 In case of doubt about whether a statement of responsibility applies to all editions or only to some, or if there is no edition statement, give such a statement in the title and statement of responsibility area. When describing the first edition, give all statements of responsibility in the title and statement of responsibility area (see 1F).

(1.2C2)

■APPLICATION   If an item has parallel edition statements but a statement of responsibility relating to the edition in only one language or script, give the statement of responsibility after all the parallel edition statements.
Optional addition. If an item has parallel edition statements and a statement of responsibility relating to the edition in more than one language or script, give each statement after the edition statement to which it relates.

☐POLICIES

National Map Collection, PAC
Give parallel statements of responsibility relating to the edition if they are in English and/or French.

2D Subsequent edition statement (3.2D)

2D1 If the item is a designated revision of a particular edition, containing changes from that edition, give the subsequent edition statement as instructed in 2D2.

9th ed., Repr. with summary of the 1961 census and supplement of additional names and amendments

4th ed., Roads rev.

₍Ed.₎ A, ₍Bar, bar, star, bar₎

8ᵃ ed., ristampa aggiornata 1977

Do not record statements relating to reissues that contain no changes unless the item is considered to be of particular bibliographic importance to the cataloguing agency.

(3.2D1)

2D2 If an item is designated as a reissue containing changes from a particular edition, give that statement following the edition statement and its statement of responsibility.

(1.2D1)

The Times atlas of the world ₍GMD₎. — Comprehensive ed., 5th ed., repr. with revisions / maps prepared by John Bartholomew & Son

₍1. Aufl. der erw. Neubearbeitung. 33. Aufl. des JRO Taschen Weltatlas₎
Note: ''Eine Mini-Ausgabe des Grossen JRO Weltatlas''

■APPLICATION   If a subsequent edition statement appears in more than one language or script, record the statement that is in the language or script of the title proper. If this criterion does not apply, record the statement that appears first. *Optionally,* give the parallel statement(s), each preceded by an equals sign.

□POLICIES
> National Map Collection, PAC
> Give parallel statements if they are in English and/or French.

2E Statements of responsibility relating to a subsequent edition statement

(3.2E)

2E1 Record a statement of responsibility relating to one or more designated subsequent editions (but not to all subsequent editions) of a particular edition following the subsequent edition statement. Follow the instructions in 1F for the transcription and punctuation of such statements of responsibility.   (1.2E1)

The elements of style ₍GMD₎ / by William Strunk, Jr. — Rev. ed. / with revisions, an introduction, and a chapter on writing by E.B. White, 2nd ed. / with the assistance of Eleanor Gould Packard

■APPLICATION   If an item has parallel subsequent edition statements but a statement of responsibility relating to the subsequent edition in only one language, give the statement of responsibility after all the parallel subsequent edition statements.
Optional addition. If an item has parallel subsequent edition statements and a statement of responsibility relating to the subsequent edition in more than one language or script, give each statement after the subsequent edition statement to which it relates.

3 *Mathematical Data Area*

National Map Collection, PAC
Give parallel statements if they are in English and/or French.

3   MATHEMATICAL DATA AREA

Contents:

3A   Preliminary rule
3B   Statement of scale
3C   Statement of projection
3D   Statement of coordinates and equinox   (3.3)

■APPLICATION   Select the method of description for cartographic items consisting of multiple maps (components; or, primary and ancillary maps), or multiple parts according to rules 0K and 1G. Based on this choice the description of the title area, mathematical data area, and physical description area must be coordinated to ensure consistency in the cataloguing record as a whole. (See Appendix G, example 35.)

This area is repeatable.

3A   Preliminary rule   (3.3A)

3A1   Punctuation. For instructions on the use of spaces before and after prescribed punctuation, see 0C.

Precede this area by a full stop, space, dash, space.

Precede the projection statement by a semicolon.

Enclose the statement of coordinates and equinox in one pair of parentheses.

If both coordinates and equinox are given, precede the statement of equinox by a semicolon.   (3.3A1)

. — Scale statement ; projection statement (coordinates ; equinox)

. — Scale statement (coordinates ; equinox)

. — Scale statement ; projection statement (coordinates)

. — Scale statement ; projection statement

. — Scale statement

. — Scale statement ; projection statement (coordinates ; equinox). — Scale statement ; projection statement (coordinates ; equinox)

. — Scale statement (coordinates ; equinox). — Scale statement (coordinates ; equinox)

. — Scale statement ; projection statement (coordinates). — Scale statement ; projection statement (coordinates)

. — Scale statement ; projection statement. — Scale statement ; projection statement

. — Scale statement . — Scale statement

50

Scale 1:7 150 000. 1 cm = 71.5 km or 1 in. = 13 miles (E 73°—E
135°/N 54°—N 18°). — Scale 1:6 000 000. 1 cm = 60 km or 1 in. =
94.7 miles ; Albers conical equal area proj., standard parallels 24° and 48°
(E 66°—E 142°/N 54°—N 14°)

3A2   Use English words and abbreviations in this area.   (3.3A2)

3B   Statement of scale   (3.3B)

■APPLICATION   When rule 1G3 is followed, and if the mathematical data for each part
are different, record the separate mathematical data statements in the same order as the
titles of the parts.
Record details relating to the grid in a note (see 7B8).

3B1a   Give the scale of a cartographic item as a representative fraction expressed as a
ratio (1:). Precede the ratio by the word *scale*. Give the scale even if it is already
recorded as part of the title proper or other title information.   (3.3B1)

Geologic map of southeast Kalimantan ₍GMD₎ = Peta Geologi
Kalimantan tenggara / compiled by the Geological Survey of Indonesia.
— Scale 1:500 000

Bartholomew one inch map of the Lake District ₍GMD₎. — Rev. —
Scale 1:63 360

Atlante automobilistico, scala 1:200 000 ₍GMD₎. — Scale 1:200 000

■APPLICATION
1. The scale must be given as a representative fraction (RF). For graphic scales and
    for maps with no scale representation, the RF must be computed. For verbal scale
    statements, the statement must be converted to a representative fraction. In both
    these cases the RF is recorded in square brackets. See Appendix B for instruc-
    tions to determine scales.
       A RF must also be recorded in square brackets for items with unfamiliar or
    obsolete units of scale.
       If a verbal scale statement appears on the cartographic item, record the state-
    ments as in 3B2.
       If only a graphic representation appears on the item, record the units used in a
    note (7B8). Where the scale or RF is not stated on the item, the method by which
    it has been computed (see Appendix B) may be given in a note (7B8). If an item
    shows any other graphic scale information, such as a composite bar scale, or a
    scale departure diagram, this information may be recorded in a note (7B8). If the
    scale is not linear, see 3B7.
       Throughout this manual, except in areas of direct transcription, e.g., title, the
    metric convention of using a space in place of a comma in the RF is followed.
    (See 0.6.)

□POLICIES
   British Library
      In the scale statement commas are not used in the RF.

3B1b *Mathematical Data Area*

Library of Congress
In the scale statement commas are used in the RF.

National Library of Australia
In the scale statement commas are not used in the RF.

National Library of New Zealand
In the scale statement commas are not used in the RF.

National Map Collection, PAC
In the scale statement commas are not used in the RF.

2. *Atlases.* For an atlas with maps at one scale only, record that scale as instructed above.

For an atlas with maps mainly at one or two scales, record the scale(s) as instructed in this rule and 3B4 respectively. In this case, there may be a few introductory or reference maps at scales different from the main scale(s). Do not include them in the scale statement, but they may be recorded in a note (7B8).

If there is no predominant scale, follow rule 3B5.

3B1b If a verbal scale statement is found on the item, record it as a representative fraction in square brackets. (3.3B1)

Scale [1:253 440]
(Verbal scale statement reads: One inch to four miles)

3B1c If a representative fraction or a verbal scale statement is found in a source other than the chief source of information, the scale is given in square brackets in the form of a representative fraction. (3.3B1)

Scale [1:63 360]

3B1d If no verbal statement of scale or representative fraction is found on the item, its container or case, or accompanying material, compute a representative fraction from a bar scale, a graticule or grid or by comparison with a map of known scale. Give the scale preceded by *ca.* (3.3B1 mod.)

Scale [ca. 1:277 740]
but
Scale ca. 1:63 360
(Scale statement as it appears on map)

■APPLICATION Enclose the computed scale and "ca." in square brackets.

3B1e If the scale cannot be determined by any of the above means, give the statement *Scale indeterminable.* (3.3B1)

3B2 *Optional addition.* Give additional scale information that is found on the item (such as a statement of comparative measures or limitation of the scale to particular parts of the item). Use standard abbreviations and numerals in place of words. Precede such additional information by a full stop.

> Scale 1:250 000. 1 in. to 3.95 miles. 1 cm to 2.5 km

Quote the additional scale information directly if (a) the statement presents unusual information that cannot be verified by the cataloguer; *or* (b) a direct quotation is more precise than a statement in conventional form; *or* (c) the statement on the item is in error or contains errors.   (3.3B2)

> Scale 1:59 403 960. ''Along meridians only, 1 inch = 936 statute miles''
>
> Scale [ca. 1:90 000] not ''1 inch to the mile''
>
> Scale [ca. 1:720 000–1:920 000]. ''Approximate vertical scale 1″ = 11.4 miles. Approximate horizontal scale 1″ = 14.5 miles''
>
> Scale ca. 1:18 000. 1,500 ft. to 1 in. and ca. 1:24 000. 2,000 ft. to 1 in.
>> (*On source:* ''Scale approx. . . . '')

■APPLICATION   For cartographic items having linear scales which are unfamiliar or no longer used, the verbal statement of scale on the item may be recorded, in quotation marks, as additional scale information, or it may be recorded in a note (see 7B8).

☐POLICIES
> British Library
>> The option is applied.
>
> Library of Congress
>> The option is applied.
>
> National Library of Australia
>> The option is applied.
>
> National Library of New Zealand
>> The option is applied.
>
> National Map Collection, PAC
>> The option is applied on a case by case basis.

3B3   If the scale within one item varies and the outside values are known, give both scales connected by a hyphen.

> Scale 1:16 000–1:28 000

If the values are not known, give the statement *Scale varies*.   (3.3B3)

■APPLICATION   This rule refers to one map/diagram/plan, etc., which is drawn on a variable scale. For example, a city map may be drawn with the central part of the city at

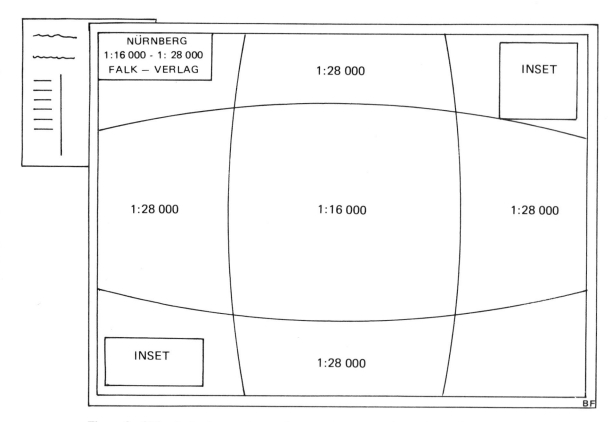

Figure 3   3B3   Scale changes at a continuous ratio, or at set intervals, out from centre
Scale 1:16 000–1:28 000

a large scale (e.g., 1:16 000) which changes at a continuous ratio outward to a smaller scale for the outskirts (e.g., 1:28 000). (See Figure 3.)

In addition, all perspective views are drawn on a variable scale, where the scale in the foreground is larger than that in the background and the scale from foreground to background changes at a continuous ratio.

For all projections the scale varies over the surface of a map according to a mathematical formula. Therefore the phrase ''scale varies'' is inapplicable to maps drawn to the specifications of a projection. For instance, a world map drawn on a Mercator projection with a given scale of 1:22 000 000 has that scale only at the equator. Although the scale at 80° north or south is 1:33 000 000, the scale at the equator, 1:22 000 000, is recorded in the scale statement. Any further information on scale may be recorded in a note (see 7B8 and Appendix B).

3B4   If the description concerns a multipart item with two scales, give both. Give the larger scale first.   (3.3B4)

Scale 1:100 000 and 1:200 000

■APPLICATION   For cartographic items lacking collective titles, but consisting of two components, or equal parts, transcribe the scales of the individual components or parts (if different) in the same order as the individual titles of these components or parts in the title area (see 1G2 and 1G3).

3B5   If the description concerns a multipart item with three or more scales, give the statement *Scales differ*.   (3.3B5 mod.)

■APPLICATION   This includes maps in components as well as multipart items. (See Glossary.)

The individual scales may be added to the appropriate title in the description of components (see 7B18).

3B6   In describing a cartographic item in which all the main maps are of one or two scales, give the scale or both scales (in the latter case give the larger scale first). Otherwise, give the statement *Scales differ*.   (3.3B6 mod.)

3B7   Give a statement of scale for celestial charts, maps of imaginary places, views (bird's-eye views or map views) and maps with nonlinear scales only if the information appears on the item. If the item is not drawn to scale, give the statement *Not drawn to scale*.   (3.3B7)

> East-west street scale approx. double to north-south scale 1 in. = 10 leagues of love
>> *(Scale given on a map of an imaginary place)*

■APPLICATION   For celestial charts compute a scale where appropriate and express it as an angular scale in mm per degree.

> Scale 88 mm per 1°

Examples of nonlinear scales are time scales, area by population density, etc.

3B8   In describing a relief model or other three-dimensional item, give the vertical scale (specified as such) after the horizontal scale if the vertical scale can be ascertained.   (3.3B8)

> Scale 1:744 080. 1 in. ca. 28 miles. Vertical scale ca. 1:96 000
>
> Scale 1:31 000 000. Vertical scale exaggerated
>
> Scale 1:17 000 000. Vertical scale 18× the horizontal scale
>
> Scale 1:250 000. Vertical scale 1:125 000. Vertical exaggeration 2:1

3C   Statement of projection   (3.3C)

3C1   Give the statement of projection if it is found on the item, its container or case, or accompanying printed material. Use standard abbreviations (see Appendix H) and numerals in place of words (see Appendix J).   (3.3C1)

> conic equidistant proj.

3C2 *Mathematical Data Area*

■ APPLICATION

1. There may be cases in which the projection is not given on the item, its container or case, or accompanying printed material, but it is known. This information may be added in square brackets in the mathematical data area.

   The peculiarities of an undetermined projection may be described in a note (7B8).

2. *Atlases.* If all of the maps (except reference or introductory maps) are drawn on the same projection, record the projection as instructed above.

   If the maps are drawn on two projections, both projections may be recorded connected by the word "and."

   If more than two projections are used, do not include a statement of projection. If considered important, they may be recorded in a note.

3C2 *Optional addition.* Add associated phrases connected with the projection statement if they are found on the item, its container or case, or accompanying printed material. Such associated phrases concern, for example, meridians, parallels, and/or ellipsoid.   (3.3C2)

> azimuthal equidistant proj. centred on Nicosia, N 35°10′, E33°22′
>
> Lambert conformal conic proj. based on standard parallels 33° and 45°
>
> transverse Mercator proj., central meridian 35°13′30″E

■ APPLICATION   Statements relating to the ellipsoid are not considered as additional information on the projection; they may be recorded in a note (see 7B8).

> Gauss proj.
> *Note:* International ellipsoid

□ POLICIES

> British Library
>   The option is applied.
>
> Library of Congress
>   The option is applied.
>
> National Library of Australia
>   The option is applied on a case by case basis.
>
> National Library of New Zealand
>   The option is applied.
>
> National Map Collection, PAC
>   The option is applied.

3D *Optional addition.* Statement of coordinates and equinox   (3.3D)

■ APPLICATION   Do not apply this option to cartographic items portraying imaginary places.

Do not use square brackets when supplying coordinates.

This area is repeatable. However, coordinates are only recorded for each title given in the title area (see 1G3). Coordinates for insets are not recorded here but may be mentioned in a note.

☐ POLICIES

    British Library
        The option is applied on a case by case basis.

    Library of Congress
        The option is applied when the information is readily determinable.

    National Library of Australia
        The option is applied when the information is readily determinable.

    National Library of New Zealand
        The option is applied to items covering an area greater than one degree.

    National Map Collection, PAC
        The option is applied in most cases. It is not applied for maps of imaginary places, nor is it applied for architectural plans for buildings that were never built. For the location of a building, only one longitudinal and one latitudinal reading is given. The option is applied only to atlases covering country level and below (e.g., national atlases, regional atlases (regions within a country), and municipal atlases).

3D1a   Give the coordinates in the following order:

        westernmost extent of area covered by item (longitude)
        easternmost extent of area covered by item (longitude)
        northernmost extent of area covered by item (latitude)
        southernmost extent of area covered by item (latitude)

                (3.3D1)

3D1b   Express the coordinates in degrees (°), minutes ('), and seconds (") of the sexagesimal system (360° circle) taken from the Greenwich prime meridian. (See also 3D1c Application 9.) Precede each coordinate by W, E, N, or S, as appropriate. Separate the two sets of latitude and longitude by a diagonal slash, and separate each longitude or latitude from its counterpart by a dash.   (3.3D1)

    (E 79°—E 86°/N 20°—N 12°)

    (E 15°00'00"—E 17°30'45"/N 1°30'12"—S 2°30'35")

    (W 74°50'—W 74°40'/N 45°05'—N 45°00')

■ APPLICATION   To convert from the centesimal system to the sexagesimal system, see Appendix B.3 Table 5.

Figure 4 may be used as a guide for recording coordinates for specific scale ranges. The boundaries between large-scale and medium-scale cartographic items and between medium-scale and small-scale cartographic items are not explicitly indicated as the

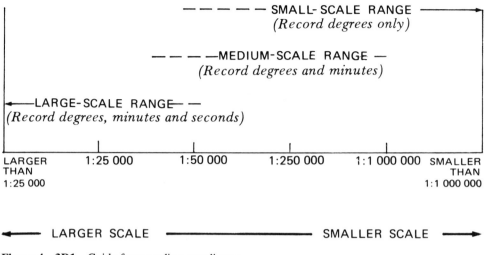

Figure 4 **3D1** Guide for recording coordinates

interpretation of such scale ranges is not standardized. The leeway shown in the diagram provides scope for assessing the case in hand.

If the coordinates given on the cartographic item are more detailed than those suggested by the above guidelines and the diagram for scale, record them as given on the cartographic item.

When the coordinates are not given on the cartographic item, establish them as accurately as possible (within the guidelines set for the scale range) by comparison to another cartographic item with coordinates that cover the same area.

Be consistent in transcribing coordinates in: degrees only; degrees and minutes only; or degrees, minutes and seconds. Supply zeros where necessary to ensure consistency.

■APPLICATION FOR SPECIFIC CASES
1. Give the longitudinal coordinates for world maps and globes as W 180°—E 180° disregarding all overlap, and disregarding the relative location of the continents. This information may be given in a note (7B8), e.g., North America at centre. Give the latitudinal coordinates as N 90°—S 90° disregarding the overlap. If the latitudinal coordinates do not extend to the poles, record the maximum extent.

    a)   (W 180°—E 180°/N 90°—S 90°)
    b)   (W 180°—E 180°/N 84°—S 70°)
    not (E 90°—E 95°/N 84°—S 70°)
         (These are the coordinates at the four corners of the map which has an overlap of 5°; North America in centre.)

    c)   (W 180°—E 180°/N 86°—S 68°)
    not (E 85°—E 85°/N 86°—S 68°)
         (These are the coordinates at the four corners of the map which has no overlap; North America in centre.)

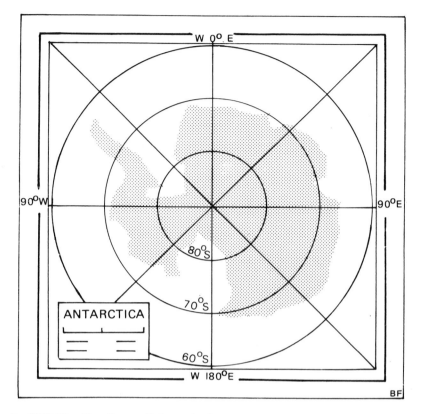

Figure 5 **3D1b(2)** Coordinates. Polar projection
(W 180°—E 180°/S 60°—S 90°)

2. Give coordinates for maps on a *polar projection* centred on the North or South Pole in the conventional form specified above. (See Figure 5.)

> (W 180°—E 180°/N 90°—N 40°)
> *(centred on the North Pole)*

> (W 180°—E 180°/S 35°—S 90°)
> *(centred on the South Pole)*

> (W 180°—E 180°/S 20°—S 90°)
> *(centred on the South Pole)*

> (W 180°—E 180°/S 60°—S 90°)
> *(centred on the South Pole)*

3. For items where the surrounding area is not treated in the same degree of detail, record the coordinates
   a)   for the four corners
   or
   b)   for the significant or principal mapped area. (See Figure 6.)

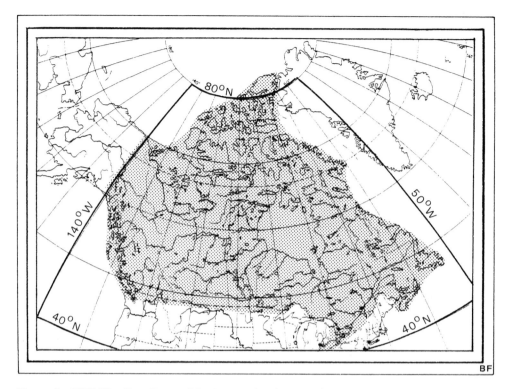

Figure 6   3D1b(3)   Coordinates. Map has varying degrees of detail
(W 141°—W 50°/N 85°—N 42°)

4. When the coordinates printed on the item do not extend to the neat line or the edge of the item, establish them as precisely as possible by extrapolation. (See Figure 7.)

5. For a cartographic item within which the scale varies (as in 3B3) record the coordinates for the maximum extent. For a multipart item where maps depict the same geographical area but are at different scales, record the coordinates for the maximum extent.

6. If the map has an inset showing a continuation of the named area which is not included in the main map, include the continuation in the statement of coordinates regardless of the scale of the extension. Any differences in scale may be recorded in a note. (See Figure 8.)

    If the map is printed in sections on one sheet, record the coordinates for the whole map as if it were joined. (See Figures 9 and 10.)

    If the map is printed on separate sheets which are meant to be joined together, record the coordinates of the whole map as if it were joined. (See Figure 11.)

7. For large-scale maps and plans depicting features such as individual buildings, glaciers, hills, etc., only one latitudinal coordinate and one longitudinal coordinate (near the centre) need be recorded.

    Also, for large-scale items which do not have coordinates and for which it is

60

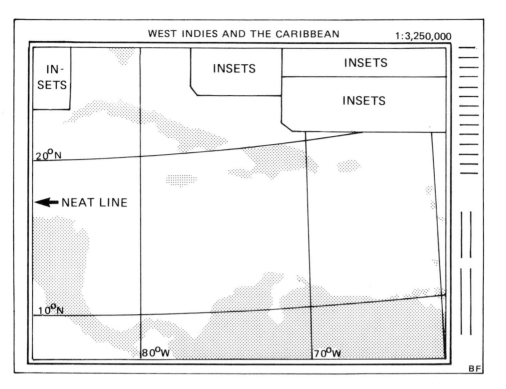

Figure 7 3D1b(4) Coordinates, extrapolated to neat line

(W 86°—W 60°/N 27°—N 7°)

not possible to establish them by comparison with a map that includes coordinates, the latitude and longitude as given in a gazetteer may be recorded. For example, the coordinates for the Bugaboo Glacier, British Columbia are:

(W 116°45′/N 50°40′)

8. For early cartographic items where the coordinates are distorted or inaccurate, record the present-day coordinates for that area.

9. When the item is of a non-terrestrial area, e.g., Moon, Mars, express the coordinates taken from the local meridian. Record the name of the meridian, if known, in a note.

10. *Atlases.* For those institutions desiring to apply this option to its records for atlases, record the coordinates for the maximum extent of the area covered by the atlas in the statement of coordinates, exclusive of reference maps.

3D1c *Optionally,* give other meridians (prime, local, or source) found on the item in the note area (see 7B8).   (3.3D1 mod.)

☐ POLICIES

British Library

The option is applied.

61

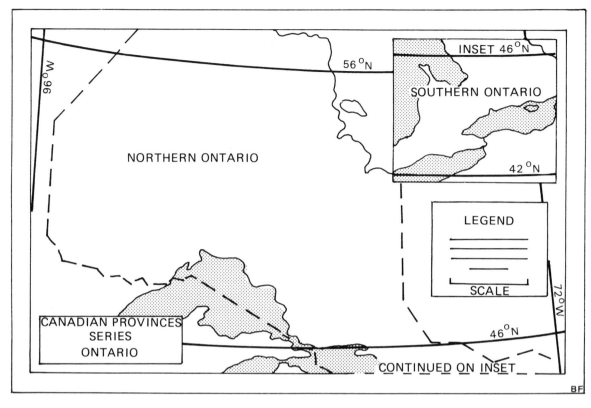

Figure 8   3D1b(6)   Coordinates. One map, continuing in an inset

(W 96°—W 72°/N 57°—N 42°)

Library of Congress
  The option is applied.

National Library of Australia
  The option is applied.

National Library of New Zealand
  The option is applied.

National Map Collection, PAC
  The option is applied.

3D2[9]   For celestial charts give as coordinates the right ascension of the item, or the right ascensions of the western and eastern limits of its collective coverage, and the

---

9. This text is a rule modification, but it is to be presented to the 2JSCAACR as a rule revision.

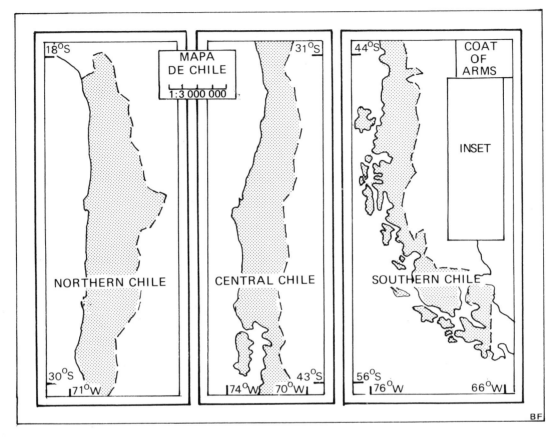

Figure 9 **3D1b(6)** Coordinates. One map printed in segments on one sheet

(W 76°—W 66°/S 18°—S 56°)

declination of the centre of the item, or the northern and southern limits of its collective coverage.

Designate the right ascension by *RA*, followed by the hours, and, where necessary, minutes and seconds of the twenty-four hour clock.

Designate the declination by *Decl.* followed by the degrees (°) and, where necessary, minutes (') and seconds (") of the sexagesimal system (360° circle), using a plus sign (+) for the northern celestial hemisphere and a minus sign (−) for the southern celestial hemisphere.

Separate right ascensions and declinations from each other by a diagonal slash, not preceded or followed by a space. Where two right ascensions are given, record both separated by the word *to*. Where two declinations are given, record both separated by the word *to*.

For celestial charts, always give the statement of equinox if coordinates are given. Express the equinox by a year preceded by a semicolon and the abbreviation *eq*. Add the

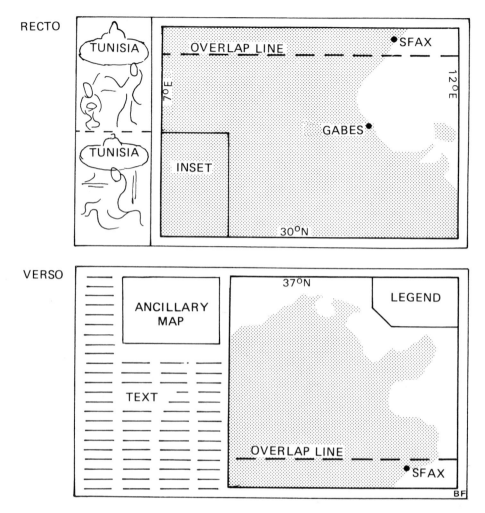

Figure 10   3D1b(6)   Coordinates. One map on both sides of one sheet
(E 7°—E 12°/N 37°—N 30°)

epoch, preceded by a comma, and the word *epoch* when it is known to differ from the equinox.

(RA 16 hr. 30 min. to 19 hr. 30 min./Decl. −16° to −49° ; eq. 1950, epoch 1948.5)

(RA 16 hr./Decl. −23° ; eq. 1950)

(RA 2 hr./Decl. +30° ; eq. 1950)

(RA 2 hr. 00 min. to 2 hr. 30 min./Decl. −30° to −45° ; eq. 1950)

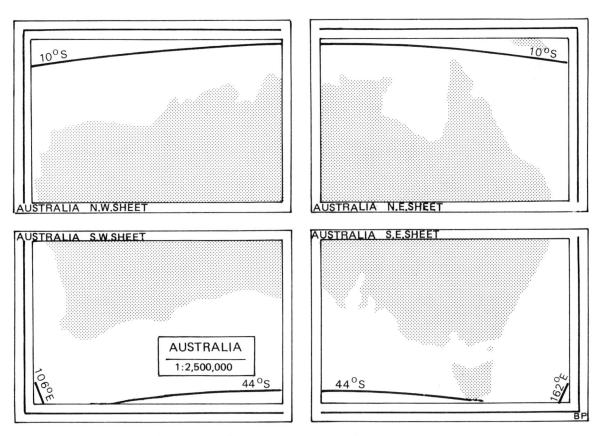

Figure 11 **3D1b(6)** Coordinates. One map on two or more sheets

(E 106°—E 162°/S 10°—S 44°)

For charts centred on a pole, indicate the declination limit.

(Centred at South Pole/Decl. limit −60°)

For atlases or collections of numerous charts arranged in declination zones, give the declination limits of each zone, but omit the statement of right ascension because it varies for each zone. If the zones are numerous, record the first few zones in order to establish the pattern, and the last zone. Indicate omissions by the mark of omission.

(3.3D2 mod.)

(Zones +90° to +81°, +81° to +63°, +63° to +45° ; eq. 1950)

(Zones +90° to +81°, +81° to +63°, . . . −81° to −90° ; eq. 1950)

3D3 Use the preceding rules and instructions to record the mathematical data of a cartographic item in microform. (11.3A mod.)

65

4 PUBLICATION, DISTRIBUTION, ETC., AREA

Contents:

4A Preliminary rule
4B General rules
4C Place of publication, distribution, etc.
4D Name of publisher, distributor, etc.
4E Statement of function of publisher, distributor, etc.
4F Date of publication, distribution, etc.
4G Place of printing, etc., date of printing, etc.   (3.4)

4A Preliminary rule   (3.4A)

4A1 Punctuation

For instructions on the use of spaces before and after prescribed punctuation, see 0C.
Precede this area by a full stop, space, dash, space.
Precede a second or subsequently named place of publication, distribution, etc., by a semicolon.
Precede the name of a publisher, distributor, etc., by a colon.
Enclose a supplied statement of function of a publisher, distributor, etc., in square brackets.
Precede the date of publication, distribution, etc., by a comma.
Enclose the details of printing, etc. (place, name, date) in parentheses.
Precede the name of a printer, etc., by a colon.
Precede the date of printing, etc., by a comma.   (3.4A1)

. — Place of publication : name of publisher, date

. — Place of publication : name of publisher, date (place of printing : name of printer, date of printing)

. — Place of publication ; place of publication : name of publisher, date

. — Place of publication : name of publisher : name of publisher, date

. — Place of distribution : name of distributor [statement of function of distributor], date

. — Place of publication : name of publisher ; place of publication : name of publisher, date

4A2 Sources of information.
Record in this area information taken from the chief source of information or from any other source specified for this area in the following rules.
Enclose information supplied from any other source in square brackets.

(1.4A2)

4B General rules   (3.4B)

4B1 This area is used to record all information about the place, name, and date of all types of publishing, distributing, releasing, and issuing activities.   (1.4B1)

4B2 Information relating to the manufacture of the item is also recorded in this area.

(1.4B2)

4B3 The places, names, and dates relating to publication, distribution, etc., may be repeated in any order that is appropriate to the item being described (see 4B8).

(1.4B3)

4B4 Give names of places, persons, or bodies as they appear, omitting accompanying prepositions unless case endings would be affected. (1.4B4)

> Berolini

> Den Haag

> . . . : Im Deutschen Verlag

but Paris

not A Paris

4B5 If the publication, distribution, etc., details appear in more than one language or script, record the details that are in the language or script of the title proper. If this criterion does not apply, record the details that appear first. (1.4B5)

■APPLICATION Give the parallel statement in a note, if desired. (See also 4D1 Application.)

4B6 If the original publication details are covered by a label containing publication details relating to a reproduction, give the publication details of the reproduction in this area. Give the publication details of the original in a note if they can be easily ascertained. (1.4B6)

■APPLICATION The rule is concerned only with labels associated with republication or reproduction, and not with a new retail outlet.

If any element of the publication, etc., area is on a stamp or label, transcribe it. If the label or stamp is not on the chief source of information, record it in square brackets (see 4A2). Make a note to convey the fact that such information is stamped on the item or expressed on a label.

The publication and distribution statement for any overprinted or revised information is recorded in the publication and distribution area. Publication information about the base map, if any, may be recorded in a note.

4B7 If an item is known to have fictitious publication, distribution, etc., details, record these in the conventional order. Supply the real publication, distribution, etc., details as a correction if these are known. (1.4B7)

> Belfast ₁i.e. Dublin₁

> Paris : Impr. Vincent, 1798 ₁i.e. Bruxelles : Moens, 1883₁

■APPLICATION *Early cartographic items.* Indicate in the note area the source of supplied publication, etc., data. When details about the publication, etc., are speculative, give them, along with any explanation, in the note area.

4B8 If an item has two or more places of publication, distribution, etc., and/or names of publishers, distributors, etc., named in it, describe it in terms of the first named place of publication, distribution, etc., and the corresponding publisher, distributor, etc. Always add the place and name of a publisher if the first named place refers to a distributor, releasing agency, etc. If any subsequent place or name is distinguished by the layout of the source of information as being that of the principal publisher, etc., add that place and name.

Follow this rule for items issued in more than one physical part when the place of publication, etc., and/or the name of the publisher, etc., changes in the course of publication.

If a place in the country of the cataloguing agency, with or without a corresponding publisher, etc., is named in a secondary position, add the information relating to it.

(1.4B8)

> London ; New York : Longmans, Green
> *(For a cataloguing agency in the United States)*

> Paris : Gauthier-Villars ; Chicago : University of Chicago Press
> *(For a cataloguing agency in the United States)*

> New York ; London : McGraw-Hill
> *(For a cataloguing agency in the United Kingdom)*

> New York : Dutton ; Toronto : Clarke, Irwin
> *(For a cataloguing agency in Canada)*

> London : Macmillan for the University of York

4B9 Early cartographic items. If the item lacks a publisher statement, give the details of the bookseller, bookseller-printer, or printer as a publisher statement if it appears that this person acted as publisher, seller, or distributor of the item.

(2.16A, 3.4B2)

4C Place of publication, distribution, etc. (3.4C)

4C1 Record the place of publication, etc., in the form and the grammatical case in which it appears. (1.4C1)

> Köln

> Lugduni Batavorum

> Canada

4C2 Add another form of the name of the place if such an addition is considered desirable as an aid to identifying the place. (1.4C2)

> Lerpwl [Liverpool]

> Christiania [Oslo]

4C3   Add the name of the country, state, province, etc., to the name of the place if it is considered necessary for identification, or if it is considered necessary to distinguish the place from others of the same name. Use the English form of the name if there is one (see *AACR2* 23.2A). Use abbreviations appearing in Appendix H.   (1.4C3)

> *City alone appears in prescribed source of information*
> Waco ₍Tex.₎
>
> London ₍Ont.₎
>
> Santiago ₍Chile₎
>
> Renens ₍Switzerland₎
>
> Lincoln ₍Neb.₎

> *City and country, etc., appear in prescribed source of information*
> Tolworth, England
>
> Carbondale, Ill.
>
> San José, Calif.
>
> San Jose, CA
>
> Milton, Qld.
>
> Burnie, Tas.
>
> Hilo, T.H.

4C4   If a place name is found only in an abbreviated form in the prescribed source of information, give it as found, and add the full form or complete the name.

(1.4C4)

> Mpls ₍i.e. Minneapolis₎
>
> Rio ₍de Janeiro₎

4C5   If a publisher, distributor, etc., has offices in more than one place and these are named in the item, always give the first named place, and the first of any subsequently named places that is in the home country of the cataloguing agency or is given prominence by the layout of the source of information. Omit all other places.   (1.4C5)

> London ; New York
> *(For a cataloguing agency in the United States)*
>
> Montréal ; Toronto
> *(Toronto given prominence by typography)*

4C6   If the place of publication, distribution, etc., is uncertain, give the probable place in the language of the chief source of information, with a question mark.

> ₍Hamburg?₎

If no probable place can be given, give the name of the country, state, province, etc.

If, in such a case, the country, state, province, etc., is not certain, give it with a question mark.

[Canada]

[Chile?]

If no probable place can be given, give the abbreviation *s.l.* (sine loco) or its equivalent in nonroman scripts. (1.4C6)

[S.l.]

4C7 *Optionally,* add the full address of a publisher, distributor, etc., to the name of the place. Enclose such an addition in parentheses. Do not add the full address for major trade publishers. (1.4C7)

London (35 Notting Hill Gate, London, W.11)

□POLICIES

British Library
The option is applied only for records included in the UK MARC data base when the publisher's address is not readily available elsewhere.

Library of Congress
The option is not applied. LC will consistently transcribe full name and address information for publishers and distributors of cartographic materials in a separate field of the machine-readable record whenever such information appears on the item or accompanying material.

National Library of Australia
ANB includes this information for minor publications not listed in the publishers' directory at the back of the four-monthly and annual cumulations. The option is not applied in *Australian Maps*.

National Library of New Zealand
The option is not applied. Addresses of occasional publishers will be supplied in parentheses following the price.

National Map Collection, PAC
The option is applied on a case by case basis.

Early cartographic items

4C8 Give the place of publication, etc., as it is found in the item. Add the modern name of the place if it is considered necessary for identification. (2.16B)

Augustae Treverorum [Trier]

■APPLICATION Give the place of publication in the orthographic form and the grammatical case in which it appears in the source of information used. If the place of publication appears together with the name of a larger jurisdiction (e.g., country, state, or similar designation), transcribe this as well.

> Elizabeth-Town

> Köln

> Apud inclytam Germaniae Basileam

> Commonwealth of Massachusetts, Boston

4C9   If the full address or the sign of the publisher, etc., appears in the prescribed sources of information, add it to the place if it aids in identifying or dating the item.
(2.16C)

> Augsburg, in S. Katharinen Gassen

> London, Fleete Streate at the signe of the Blacke Elephant

4C10   Supply the name of the place of publication in English (in square brackets) when only an address or sign appears in the publication. (Record the address or sign within the publisher statement.) When supplying the place, give a justification in the note area if necessary.
(2.16D mod.)

> [Paris]
> (*Imprint reads:* "à l'enseigne de l'éléphant," the trade sign of a Paris printer)

> [London]
> (*Imprint reads:* "sold in St. Paul's Church yard")

> [Cambridge, Mass.] : Printed by Samuel Green, 1668
> *Note:* The printer, Samuel Green, was located in Cambridge, Mass.,
> from 1660 to 1672

4C11   If more than one place of publication, etc., is found in the item, always record the first, and *optionally*, record the others in the order in which they appear. If second or subsequent places are omitted, add [etc.].
(2.16E)

> Londres ; et se trouve à Paris

> London [etc.]

□ POLICIES
> British Library
>> The "optionally" provision is applied.

> Library of Congress
>> The "optionally" provision is applied.

> National Library of Australia
>> The "optionally" provision is applied.

National Library of New Zealand
The "optionally" provision is applied.

National Map Collection, PAC
The "optionally" provision is applied.

4D   Name of publisher, distributor, etc.   (3.4D)

■APPLICATION   The following are some terms which indicate publishing, manufacturing, issuing, or printing functions.

> issued *or* reissued
> made
> ₁photo₁ composed
> printed *or* reprinted
> produced *or* reproduced
> published *or* republished
> ₁all commercial methods of production or reproduction; e.g.,
>   lithographed, zincographed, offset₁

For early printed cartographic items see 4B9, 4D9, 4D10.

4D1   Give the name of the publisher, distributor, etc., following the place(s) to which it relates.   (1.4D1)

University, Ala. : Geological Survey of Alabama

Dar es Salaam : Surveys and Mapping Division, Ministry of Lands, Housing and Urban Development

Austin : Bureau of Economic Geology, University of Texas at Austin

Berlin ; Stuttgart : Reise- und Verkehrsverlag ; Stuttgart : Vertrieb, Geo Center

San Jose, Calif. : Goushā/Chek-Chart ; North Hollywood, Calif. : Exclusive distribution in Southern California by Mitock Publishers, c1978

■APPLICATION   If the name of the publisher, distributor, etc., appears in more than one language or script, record the name that is in the language or script of the title proper. If this criterion does not apply, record the name that appears first.

Optionally, give the parallel names, each preceded by an equals sign. (See also 4B5 Application.)

□POLICIES

National Map Collection, PAC
Give parallel names that appear in English and/or French for the first named publisher, distributor, etc., any prominently named publisher, distributor, etc., and the first named Canadian publisher, distributor, etc. For rare items give parallel names that appear in English and/or French for all publishers, distributors, etc.

4D2   Record the name of the publisher, etc., and *optionally* the distributor as instructed in the following rules.   (3.4D1)

> Southampton : Ordnance Survey

> Point Reyes, Calif. : Drake Navigators Guild

> Paris : Institut géographique national

> ₁London₁ : Royal Geographical Society

> Montréal : Éditions FM

> ₁Chicago₁ : Chicago Area Transportation Study

> Amsterdam ; London : North-Holland Pub. Co.
> *(For a cataloguing agency in the United Kingdom)*

> London : Royal Geographical Society ; Lympne Castle, Kent : H. Margary
> *(Second publisher given prominence by layout)*

> Southampton : Ordnance Survey for the Institute of Geological Sciences

> Tananarive : Service géographique de Madagascar

■ APPLICATION   If only the address or a device of the publisher, etc., appears in the prescribed sources of information, supply the name of the publisher, etc., in square brackets, if known, either before or after the address or device, as appropriate. Whenever the name of a publisher is supplied, supporting evidence may be given in a note.

☐ POLICIES

> British Library
> The option is applied.

> Library of Congress
> The option is applied.

> National Library of Australia
> The option is applied on a case by case basis.

> National Library of New Zealand
> The option is applied.

> National Map Collection, PAC
> The option is applied on a case by case basis.

4D3   Give the name of a publisher, distributor, etc., in the shortest form in which it can be understood and identified internationally.   (1.4D2)

> : Penguin

> *not*   : Penguin Books

    : W.H. Allen

   not : Allen
           (Avoids confusion with other publishers called Allen)

4D4   Do not omit from the phrase naming a publisher, distributor, etc.:
   a)   words or phrases indicating the function (other than solely publishing) performed
        by the person or body

        : Map Productions ₍for the₎ Royal Automobile Club

        : For sale by the Supt. of Docs., U.S. G.P.O.

        : Distributed by Denoyer-Geppert Co.

        : Available from Intergovernmental Oceanographic Commission

        : ₍Distribución exclusiva para librerías, Editorial Cantábrica, 1976?₎

   but : Published for the Social Science Research Council by Heinmann

   not : For the Social Research Council by Heinmann

   b)   parts of the name required to differentiate between publishers, distributors, etc.
                                                                              (1.4D3)

        : Longmans, Green

        : Longmans Educational

   not : Longmans

4D5   If the name of the publisher, distributor, etc., appears in a recognizable form in
the title and statement of responsibility area, give it in the publication, distribution, etc.,
area in a shortened form. If, in such a case, the publisher, distributor, etc., is a person
rather than a corporate body, give the initials and the surname of the person.
                                                                              (1.4D4)

        Atlante internazionale ₍GMD₎ : del Touring club italiano. — 8ª ed.,
        ristampa aggiornata 1977. — Scales differ. — Milano : Lo Club, 1978

        Atlas do estado da Bahia ₍GMD₎ / Governo do Estado, Secretaria do
        Planejamento, Ciência e Tecnologia, Centro de Planejamento da Bahia.
        — Scale 1:2 500 000. — Salvador ₍Brazil₎ : O Centro, 1976–

        List grosser Weltatlas ₍GMD₎ : Mensch und Erde / ₍Gesamtausführung,
        Paul List Verlag, München₎. — Ausg. Rheinland-Pfalz und Saarland, 2.
        Aufl. / ₍Redaktion, Volkhard Binder₎. — Scales differ. — München :
        List, 1976, c1975

        Ludington, Mich. ₍GMD₎ : 1892 / drawn and published by C.J. Pauli.
        — Not drawn to scale. — Milwaukee, Wis. : C.J. Pauli, ₍1892₎

        Topograficzna karta Królestwa Polskiego ₍GMD₎ : (1822–1843) /
        Bogusław Krassowski ; Biblioteka Narodowa, Zakład Zbiorów
        Kartograficznych. — Scale 1:126 000. — Warszawa : BN, 1978

but

>
> Atlas de Tunisie ₁GMD₁ / sous la direction de Mohamed Fakhfakh et
> sous le patronage de Georges Laclavère ; préf. de Mohamed Mzali. —
> Paris : Éditions J.A. ₁Jeune Afrique₁, c1979

4D6 If two or more agencies are named as performing the same function, always
include the first named agency and add any agency given prominence by typography.

(1.4D5)

>
> Toronto : McClelland and Stewart : World Crafts Council
> *(Second publisher given prominence by typography)*
>
> ₁London₁ : Oxford University Press in association with Geoprojects

■APPLICATION   The cartographic design of the item in question may not always make
clear the relative importance of the various publishers, distributors, etc. Give first the
name emphasized by its position relative to the other names, or by typography; if this
cannot be applied, select the first name reading from top to bottom, and from left to
right, where appropriate.

If, however, the names of publishers or distributors do not appear sequentially on the
cartographic item, record first the name of the publisher or distributor considered most
important by the cataloguing agency. The other names may also be recorded.

4D7 If the name of the publisher, distributor, etc., is unknown, give the abbreviation
s.n. (sine nomine) or its equivalent in nonroman scripts.   (1.4D6)

>
> Paris : ₁s.n.₁
>
> Sherbrooke, Québec : ₁s.n.₁

4D8 In case of doubt about whether a named agency is a publisher or a manufacturer,
treat it as a publisher.   (1.4D7)

Early cartographic items

4D9 Record the rest of the details relating to the publisher, etc., as they are given in
the item. Separate the parts of a complex publisher, etc., statement only if they are
presented separately in the item. If the publisher, etc., statement includes the name of a
printer, record it here. Omit words in the publisher, etc., statement that do not aid in the
identification of the item and do not indicate the role of the publisher, etc. Indicate
omissions by the mark of omission.   (2.16F)

>
> London : R. Barker
>
> London : Printed for the author and sold by J. Roberts
>
> London : Imprinted . . . by Robt. Barber . . . and by the assigns of
> John Bill
>
> Birmingham : Printed by John Baskerville for R. and J. Dodsley . . .
>
> Paris : Chez Testu, imprimeur-libraire
>
> Paris : Ex officina Ascensiana : Impendio Joannis Parvi

4D10 If there is more than one statement relating to publishers, etc., always record the first statement, and *optionally,* record the other statements in the order in which they appear. If subsequent statements are omitted, add ₍*etc.*₎. (2.16G)

> London : Printed for the author and sold by J. Parsons ₍etc.₎

□ POLICIES
>
> British Library
> The "optionally" provision is applied.
>
> Library of Congress
> The "optionally" provision is applied.
>
> National Library of Australia
> The "optionally" provision is applied.
>
> National Library of New Zealand
> The "optionally" provision is applied.
>
> National Map Collection, PAC
> The "optionally" provision is applied.

4E *Optional addition.* Statement of function of publisher, distributor, etc.

(3.4E)

4E1 Add to the name of a publisher, distributor, etc., one of the terms below:

> distributor
> publisher
> producer
> *(Used for producing entity other than a production company)*
> production company

unless:

   a) the phrase naming the publisher, distributor, etc., includes words that indicate the function performed by the person(s) or body (bodies) named

or b) the function of the publishing, distributing, etc., agency is clear from the context. (1.4E1)

> København : Geodaetisk Institut ; ₍London₎ : Stanford ₍distributor₎

> . . . ; London ₍Ont.₎ : Western News Co. ₍distributor₎ . . .

> London : Macmillan : Educational Service ₍distributor₎

but

> New York : Released by Beaux Arts

□ POLICIES
>
> British Library
> The option is applied.

Library of Congress
    The option is applied when the information is necessary for clarification and is readily ascertainable.

National Library of Australia
    The optional addition is exercised at the discretion of the cataloguer, and is added to clarify the function of a particular person or body where necessary.

National Library of New Zealand
    The option is applied.

National Map Collection, PAC
    The option is applied when the information is necessary for clarification and is readily ascertainable. Liberal use of the option is discouraged and no strict uniformity is expected.

4F   Date of publication, distribution, etc.   (3.4F)
    For early cartographic items, see 4F10, 4F11.

4F1   Give the date of publication, distribution, etc., of the edition named in the edition area. If there is no edition statement, give the date of the first edition. Give dates in Western-style arabic numerals. If the date found in the item is not of the Gregorian or Julian calendar, give the date as found and follow it with the year(s) of the Gregorian or Julian calendar. (See also 4F10, 4F11.)   (1.4F1)

    , 1975

    , 4308 [1975]

    , [4308 i.e. 1975]

    , 5730 [1969 or 1970]

    , anno 18 [1939] (*not* anno XVIII)

    , 1976 (*not* ı ꟼ v ꟼ)

    Budapest : Cartographia, 1979/1980
    Note: "Compilation date 15th July, 1978"— [P. 4]
        (*Imprint:* Cartographia, Budapest, 1979/1980, from preliminaries)

    Budapest : Cartographia, 1979/1980
    Note: "Compilation date 30th April, 1979"—P. iv
        (*Imprint:* Cartographia, Budapest, 1979/1980, from preliminaries)

■APPLICATION
    1. If no publication date, copyright date, or printing date is found on the item, a publication date can often be inferred from other information appearing on the cartographic item. Enclose inferred dates in square brackets. If it is doubtful that the inferred date is the actual publication date, add a question mark within the square brackets.

The date of publication can be inferred from the following sources:

a) date in the title proper, other title information, or alternate title

> *Title:* 1980 official highway map
> *Record publication date:* ₁1980₁

b) date in the statement of responsibility

> *Author statement:* prepared by the Dept. of Lands and Surveys, 1978
> *Record publication date:* ₁1978₁

c) date in the edition statement

> *Edition statement:* Rev. 1975
> *Record publication date:* ₁1975₁

> *Edition statement:* 1979– 80 ed.
> *Record publication date:* ₁1979₁

d) printing or publisher's code

> *Code on U.S. Central Intelligence Agency map:* 503821 6-78
> *Record publication date:* ₁1978₁

> *Code on United Nations map:* CART-M-74-3
> *Record publication date:* ₁1974₁

e) other information appearing elsewhere on the item

> *Statistical data in text:* 1975 estimated population
> *Record publication date:* ₁1975?₁

> *Note on map:* freeway to open in fall 1980
> *Record publication date:* ₁1980?₁

When inferring a date of publication, do not consider the following kinds of information:

a) date of geodetic control

> 1927 North American datum

b) date of magnetic declination

> 1975 magnetic north declination

c) dates of boundaries

> cease-fire lines as of 1967

> international boundaries as of Sept. 1, 1939

d) dates in base map notes

> Based on the 1972 Forest Service class A map

> Base map prepared by U.S. Geological Survey in 1969

2. If the date on an accompanying text varies from that given on the item, record the date on the item itself in the publication, distribution area.

, 1974 ₍i.e. 1975₎
Note: Maps compiled in 1975 from 1970 census data

, 1975 ₍i.e. 1977₎
Note: Date on cover label 1977
    (*Braille t.p. reads:* 1975; *typed cover label reads:* 1977)

4F2   Give the date as found in the item even if it is known to be incorrect. If a date is known to be incorrect, add the correct date.

, 1697 ₍i.e. 1967₎

If necessary, explain any discrepancy in a note.   (1.4F2)

, 1963 ₍i.e. 1971₎
Note: Originally issued as a sound disc in 1963; issued as a cassette in 1971

4F3   Give the date of a particular reissue of an edition as the date of publication only if the reissue is specified in the edition area (see 2D). In this case, give only the date of the reissue.   (1.4F3)

4F4   If the publication date differs from the date of distribution, add the date of distribution if it is considered to be significant by the cataloguing agency. If the publisher and distributor are different, give the date(s) after the name(s) to which they apply.

London : Macmillan, 1971 ₍distributed 1973₎

London : Educational Records, 1973 ; New York : Edcorp ₍distributor₎, 1975

Toronto : Royal Ontario Museum, 1971 ; Beckenham ₍Kent₎ : Edward Patterson ₍distributor₎
(*Distribution date known to be different but not recorded*)

If the publication and distribution dates are the same, give the date after the last named publisher, distributor, etc.   (1.4F4)

New York : American Broadcasting Co. ₍production company₎ : Released by Xerox Films, 1973

□POLICIES ON ADDING DATE OF DISTRIBUTION
British Library
    The option is applied on a case by case basis.

Library of Congress
    The option is applied on a case by case basis.

National Library of Australia
    The option is applied on a case by case basis.

National Library of New Zealand
The option is applied on a case by case basis.

National Map Collection, PAC
The option is applied on a case by case basis.

4F5 *Optional addition.* Add the latest date of copyright following the publication, distribution, etc., date if it is different. (1.4F5)

, 1967, c1965

■APPLICATION   When a copyright date is recorded, precede it by a lowercase ''c'' regardless of the particular symbol, abbreviation, or other designation that has been used to indicate copyright, e.g., c, cop., copyright, etc. Note that what is being expressed is the fact of copyright in a particular year, not a literal transcription from the item.

☐POLICIES
British Library
The option is applied.

Library of Congress
The option is applied.

National Library of Australia
The option is applied.

National Library of New Zealand
The option is not applied.

National Map Collection, PAC
The option is applied.

4F6 If the dates of publication, distribution, etc., are unknown, give the copyright date or, in its absence, the date of manufacture (indicated as such) in its place.

(1.4F6)

, c1967

, 1967 printing

Madrid : RENFE, c1975
    (*On source:* Dep. Legal . . . 1975; *transcription:* c1975)

Paris : Ponchet, c1975
    (*On source:* Dépôt légal . . . 1975; *transcription:* c1975)

4F7 If no date of publication, distribution, etc., copyright date, or date of manufacture can be assigned to an item, give an approximate date of publication.

, [1971 or 1972]   *One year or the other*

, [1969?]   *Probable date*

, ₁between 1906 and 1912₁	*Use only for dates less* *than 20 years apart*	
, ₁ca. 1960₁	*Approximate date*	
, ₁197–₁	*Decade certain*	
, ₁197–?₁	*Probable decade*	
, ₁18—₁	*Century certain*	
, ₁18—?₁	*Probable century*	(1.4F7)

■APPLICATION   Use publishers' lists and catalogues, and printed bibliographies if special searching for a more precise publication date is desired.

4F8   If two or more dates are found on the various parts of a multipart item (e.g., if such an item is published in parts over a number of years), give the earliest and latest dates.

, 1968–1973

In describing a multipart item that is not yet complete, give the earliest date only, and follow it with a hyphen and four spaces.

, 1968–

Optionally, when the item is complete, add the latest date.   (1.4F8)

□POLICIES
  British Library
  The option is applied.

  Library of Congress
  The option is applied.

  National Library of Australia
  The option is applied.

  National Library of New Zealand
  The option is applied.

  National Map Collection, PAC
  The option is applied.

4F9   Manuscripts. Give the date or inclusive dates of the manuscript or manuscript collection unless it is already included in the title (as with letters and legal documents). Give the date as a year or years, and *optionally* the month and day (in the case of single manuscripts), in that order.   (4.4B1 mod.)

Exil ₁GMD₁ / St.-J. Perse. — 1941

Correspondence ₁GMD₁ / William Allen. — 1821–1879

Records ₍GMD₎ / American Colonization Society. — 1816–1908

Alice's adventures under ground ₍GMD₎ : a Christmas gift to a dear child in memory of a summer day / ₍Lewis Carroll (Rev. C.L. Dodgson)₎. —1864

Sonnet, To Genevra ₍GMD₎ / ₍Lord Byron₎. — 1813 Dec. 17

☐ POLICIES

British Library
The option is not applied.

Library of Congress
The option is applied whenever the information is readily available.

National Library of Australia
The option is applied whenever the information is readily available.

National Library of New Zealand
The option is applied whenever the information is readily available.

National Map Collection, PAC
The option is applied when the month and day can be readily ascertained from the manuscript itself. When they cannot, they will be sought only when recording them would appear to be of great importance.

Early cartographic items

4F10 Give the date of publication or printing, including the day and month, as found in the item. Change roman numerals indicating the year to arabic numerals unless they are misprinted, in which case record the roman numerals and add a correction. Add the date in the modern chronology if this is considered to be necessary. (2.16H)

1716

iv Ian 1497

xii Kal. Sept. ₍21 Aug.₎ 1473

In vigilia S. Laurentii Martyris ₍9 Aug.₎ 1492

iii Mar. 1483 ₍i.e. 1484₎

1733
   (*Date in book:* MDCCXXXIII)

DMLII ₍i.e. 1552₎

Optionally, formalize the date if the statement appearing in the item is very long.
(2.16H mod.)

xviii Mai. 1507
   (*not* Anno gratiae millesimo quingentesimo septimo die vero decimoctavo Maij)

☐ POLICIES

    British Library

      1) The date in modern chronology is always added.

      2) The option is applied.

    Library of Congress

      1) The date in modern chronology is always added.

      2) The option is applied.

    National Library of Australia

      1) The date in modern chronology is always added.

      2) The option is applied.

    National Library of New Zealand

      1) The date in modern chronology is always added.

      2) The option is applied.

    National Map Collection, PAC

      1) The date in modern chronology is always added.

      2) The option is applied.

4F11   If the item is undated and the date of publication is unknown, give an approximate date. (2.16J)

    [1492?]

    [not after Aug. 21, 1492]

    [between 1711 and 1719]

4G   Place of manufacture, etc., name of manufacturer, etc., date of manufacture, etc. (1.4G)

4G1   If the name of the publisher is unknown, give the place and name of the printer or manufacturer, if they are found on the item, its container or case, or accompanying printed material. (3.4G1)

    Paris : [s.n., ca. 1898] (Paris : LeBrun)

    [S.l. : s.n., 1979] (Genève : Impr. Atar)

    [Madrid? : s.n.], 1977 (Madrid : Litograph)

    [Nieuwpoort? Belgium : s.n., 1980] (Poperinge [Belgium] :
Vansevenant)

4G2   In recording the place and name of the manufacturer, follow the instructions in 4B–4D. (1.4G2)

4G3   If the date of manufacture is given in place of an unknown date of publication, distribution, etc. (see 4F6), do not repeat it here. (1.4G3)

4G4   *Optional addition.* Give the place, name of printer, etc., and/or date of printing, etc., if they differ from the place, name of publisher, etc., and date of publication, etc.; and are found on the item, its container or case, or accompanying printed material; and are considered important by the cataloguing agency.   (3.4G2)

> Austin : Texas Dept. of Water Resources, 1977 (1978 printing)

> London : Laurie & Whittle, 1804 (1810 printing)

☐POLICIES

> British Library
> The option is applied on a case by case basis.

> Library of Congress
> The option is applied. No great effort at uniformity is attempted.

> National Library of Australia
> The option is applied at the discretion of the cataloguer.

> National Library of New Zealand
> The option is applied on a case by case basis.

> National Map Collection, PAC
> The option is applied if such information is "considered important,"
> i.e., rarely.

4G5   Early cartographic items. If the printer of an early printed atlas is named separately in the item and the function as printer can be clearly distinguished from that of the publisher or bookseller, give the place of printing and the name of the printer as instructed in the preceding rules.   (2.16K mod.)

5   PHYSICAL DESCRIPTION AREA

Contents:

■APPLICATION   Select the method of description for cartographic items consisting of multiple maps (components; or, primary and ancillary maps), or multiple parts according to rules 0K and 1G. Based on this choice the description of the title area, mathematical data area, and physical description area must be coordinated to ensure consistency in the cataloguing record as a whole. (See Appendix G, example 35.)

5A   Preliminary rule   (3.5A)

5A1   Punctuation
For instructions on the use of spaces before and after prescribed punctuation see 0C.

Precede this area by a full stop, space, dash, space *or* start a new paragraph.
Precede other physical details by a colon.
Precede dimensions by a semicolon.
Precede a statement of accompanying material by a plus sign.
Enclose physical details of accompanying material in parentheses.   (3.5A1)

   . — Specific material designation and extent of item : other physical details ; dimensions

   . — Specific material designation and extent of item : other physical details ; dimensions + accompanying material statement

   . — Specific material designation and extent of item ; dimensions

■ APPLICATION
Order of elements

          extent (number of items)
          specific material designation
          other physical details
             layout on verso and recto, i.e., *back to back* and *both sides*
               (for maps only)
             manuscript
             photocopying
             number of maps in an atlas
             colour
             material
             mounting
          dimensions
          accompanying material

5A2   Sources of information. Take information for this area from any source. Take explicitly or implicitly stated information from the item itself. Enclose information in square brackets only when specifically instructed by the following rules.   (1.5A2)

5A3   If an item is available in different formats (e.g., as text and microfilm), give the physical description of the format in hand. *Optionally,* make a note describing other formats in which it is available (see 7B16).   (1.5A3 mod.)

☐ POLICIES
   British Library
      The option is applied only for records included in the UK MARC data base.

   Library of Congress
      For cartographic materials the option is applied when ''other formats'' are known to exist at the time of cataloguing. No additional research will be undertaken on a regular basis.

   National Library of Australia
      The option is applied.

National Library of New Zealand
   The option is applied.

National Map Collection, PAC
   The option is applied when "other formats" are known to exist at the time of cataloguing. No additional research will be undertaken.

5B   Extent of item (including specific material designation)   (3.5B)

■APPLICATION   Due to the confusion in *AACR2* 3.5B concerning *parts, sheets,* and *section,* the manual uses the terms *components, sheets,* and *parts* as below.
   1.   *Components.*   A component is one of two or more maps all of equal importance appearing on a single sheet. No map is the dominant or "main" map, and they may or may not have a collective title and/or individual titles. (See Figure 12.)
   2.   *Sheets.*   One map on more than one sheet[10] (unassembled). (See Figure 13.)
   3.   *Parts.*   For cartographic material, a part is a physically separate unit that can stand alone bibliographically. It may be used in conjunction with other units of similar design, e.g., a group of maps each of a common area but displaying various themes; or, a group of maps designed primarily for use individually or to provide total coverage of a specific area. (See Figure 14.)

5B1a   Record the number of physical units of a cartographic item by giving the number of units in arabic numerals and one of the following terms, as appropriate.

aerial chart	map section
aerial remote sensing image	orthophoto
anamorphic map	photo mosaic
atlas	(controlled)
bird's-eye view *or* map view	photo mosaic
block diagram	(uncontrolled)
celestial chart	photomap
celestial globe	plan
chart	relief model
globe	remote-sensing image
(for globes other	space remote-sensing
than celestial globes)	image
hydrographic chart	terrestrial remote-
map	sensing image
map profile	(3.5B1 mod.)

   1 aerial chart

   1 atlas

   1 celestial globe

   3 plans

---

10. Sheet. As used in this area, a single piece of paper with manuscript or printed matter on one or both sides.

Figure 12

3 maps on 1 sheet . . . *(Physical description)*
Components: . . . *(Note area, 7B18)*

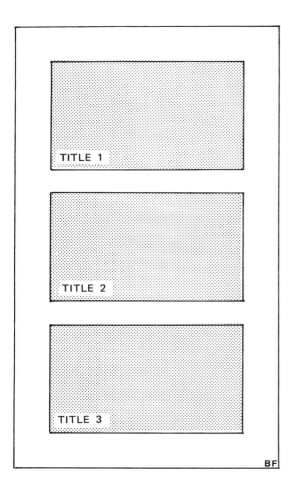

Figure 13

1 map on 2 sheets . . . *(Physical description)*

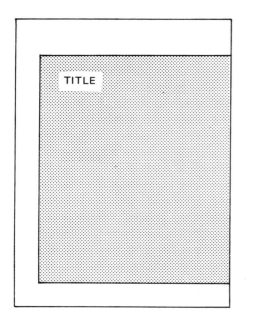

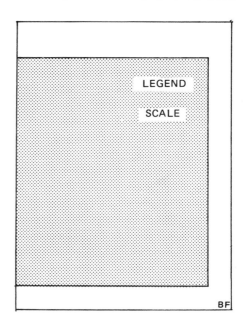

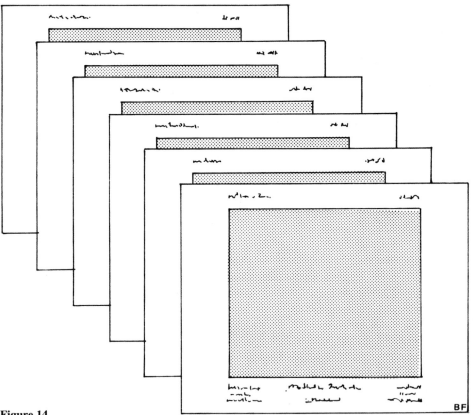

Figure 14

6 maps . . . *(Physical description)*
Parts: . . . *(Note area, 7B18)*

■APPLICATION   As an alternative to the preceding list, the number of specific material designators may be restricted to a few general terms. The following list of nine such general terms relates only to the form in which the information is presented and not to content or subject matter. Following this short list are some of the specific types of items which are subsumed by these nine general terms.

> atlas
> diagram
> globe
> map
> model
> profile
> remote-sensing image
> section
> view

atlas	atlas factice
	Bible atlas
	boundary atlas
	braille atlas
	city atlas
	county atlas
	facsimile atlas
	fire insurance atlas
	geographical atlas
	geological atlas
	historical atlas
	hydrographical atlas
	linguistic atlas
	manuscript atlas
	maritime atlas
	military atlas
	national atlas
	pilot guide
	plat book
	pocket atlas
	portolan atlas
	regional atlas
	road atlas
	school atlas
	state atlas
	statistical atlas
	thematic atlas
	topographical atlas
	wall atlas or flip chart (atlas)
diagram	block diagram
	fence diagram
	reliability diagram
	slope diagram
	triangulation diagram
globe	celestial globe
	moon globe
	terrestrial globe
	[name of planet] globe
map	aeronautical chart
	anaglyphic map
	anamorphic map
	base map
	braille map
	cadastral map
	celestial chart

	chart
	facsimile map
	historical map
	hydrographic map, chart
	index map
	key map
	layered (relief) map
	location map
	manuscript map
	map of imaginary place(s)
	orthophotomap
	outline map
	photomap
	pictorial map
	pictorial relief map
	plan
	plat
	relief map
	schematic map
	series map
	sketch map
	strip map
	tactile map
	thematic map
	topographic map
	wall map
model	planetarium (model of the solar system)
	relief model
profile	
remote-sensing image	aerial remote-sensing image
	airphoto, air photograph
	controlled photomosaic
	high oblique air photo, high oblique airphotograph
	infrared scanning image
	low oblique air photo, low oblique airphotograph
	multispectral photo image
	multispectral scanning image, (MSS) image
	orthophotograph
	orthophotomosaic
	photomosaic
	Sidelooking Airborne Radar (SLAR) image

Synthetic Aperture Radar
(SAR) image
uncontrolled photomosaic
vertical airphoto, vertical
airphotograph
₁name of satellite₁ image

section

view   bird's-eye view
                                panorama
                                panoramic drawing
                                perspective view
                                worm's-eye view

□POLICIES

British Library
The British Library uses the restricted list of SMDs. In addition, terms
from other chapters in *AACR2* are used as required.

Library of Congress
LC uses the restricted list of SMDs. In addition terms from other
chapters in *AACR2* are used as required.

National Library of Australia
The NLA uses the restricted list of SMDs. In addition, terms from other
chapters in *AACR2* are used as required.

National Library of New Zealand
The NLNZ uses the restricted list of SMDs. In addition, terms from
other chapters in *AACR2* are used as required.

National Map Collection, PAC
The NMC uses the restricted list of SMDs. In addition, terms from
other chapters in *AACR2* are used as required.

5B1b   If the parts of the item are very numerous and the exact number cannot be
readily ascertained, give an approximate number.   (3.5B1)

ca. 800 maps

■APPLICATION   Generally, prefer not to guess the number of parts. If at all possible,
count the items. If the count is not verified (i.e., counted more than once), cite the
number of pieces as approximate. If a multipart item is incomplete, see 5B28.

5B1c   If the cartographic item is not comprehended by one of the above terms, use an
appropriate term (preferably taken from rule 5B of one of the chapters of Part I of
AACR2).   (3.5B1)

52 playing cards

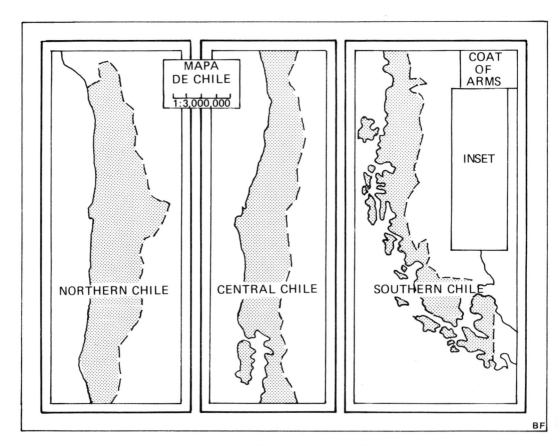

5B2a *Physical Description Area*

122 slides

   (*Title proper:* World color regions atlas)

1 jigsaw puzzle (133 pieces)

5B2a If a map is printed in segments on one sheet (see Figure 15) or on recto and verso and so designed that the segments could be fitted together to form a single map, describe it as "1 map" or as "1 map : both sides" as appropriate. If maps, plans, etc., are printed on two or more sheets but so designed that they could be fitted together to form a single map, plan, etc., or more than one map, plan, etc., give the number of complete maps, plans, etc., followed by the number of sheets. (See Figures 15, 16, and 17.)   (3.5B2 mod.)

1 map : both sides, col.

1 map on 4 sheets

Figure 15   5B2a   Extent. One map printed in segments on one sheet

1 map . . . (*Explain the layout in a note, if necessary*)

RECTO

VERSO

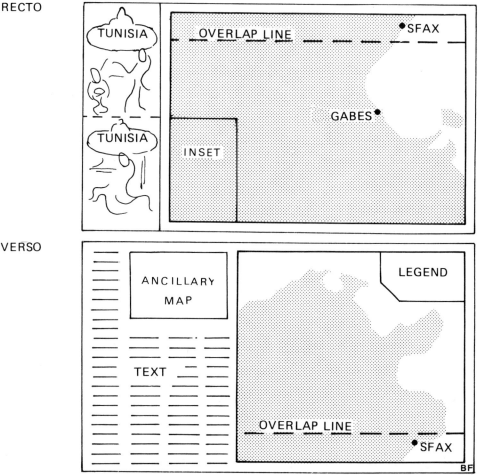

Figure 16   5B2a   Extent. One map printed on both sides of one sheet

1 map: both sides, col.

■APPLICATION   Describe the physical state of the item in hand at the time of cataloguing (see 0.4) regardless of how it was issued by the publisher/printer, e.g., if an item was originally issued on one sheet but subsequently dissected, it is described in its dissected form. If the map, plan, etc., has been issued originally in a number of sheets then subsequently assembled into one sheet, the number of sheets may be specified in a note. (See Figure 17.)

   1 map : col., dissected and mounted on linen

   1 map : col., dissected in 4 pieces
   Note: Originally printed in 2 sheets

For maps printed on both sides, see also 5C1 and 5D1.

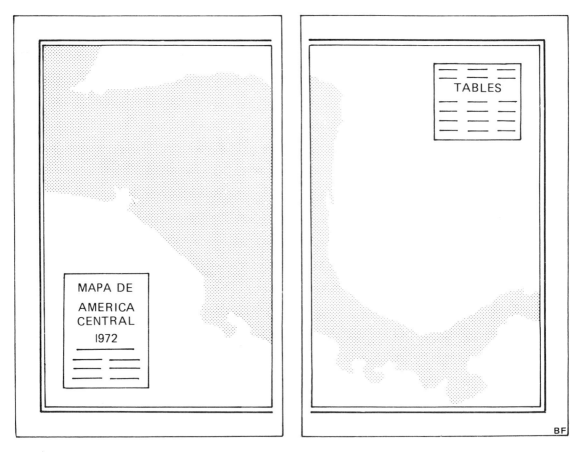

Figure 17   5B2a   Extent. One map on two or more sheets

1 map on 2 sheets

5B2b   If there is more than one map, plan, etc., on a sheet, specify the number of maps, etc. (See Figure 18.)   (3.5B2)

       4 maps on 1 sheet

■APPLICATION   This rule refers to sheets with more than one map, of approximately equal importance, and not to maps with insets.

5B2c   If the item consists of a number of sheets each of which has the characteristics of a complete map, plan, etc., treat it as a collection and describe it as instructed in 5B1.

       (3.5B2)

Physical description for atlases

5B3   Add, to the statement of extent for an atlas, the pagination or number of volumes as instructed in 5B4– 5B26.   (3.5B3)

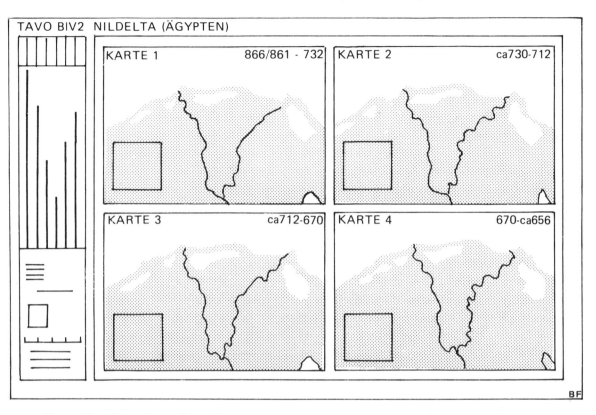

Figure 18   5B2b   Extent. More than one map on one sheet

4 maps on 1 sheet

1 atlas (3 v.)

1 atlas (xvii, 37 p., 74 leaves of plates)

Single volumes

5B4   Record the number of pages or leaves in a publication in accordance with the terminology suggested by the volume. That is, describe a volume with leaves printed on both sides in terms of pages, describe a volume with leaves printed on one side only in terms of leaves, and describe a volume that has more than one column to a page and is numbered in columns rather than pages in terms of columns. If a publication contains sequences of leaves and pages, or pages and columns, or leaves and columns, record each sequence. Describe a volume printed without numbering in terms of leaves or pages, but not of both. For the treatment of unnumbered leaves of plates, see 5B10. Describe a broadside as such. Describe a folder and other single sheets as *sheet.* Describe a case or portfolio as such. (2.5B1)

5B5   Record the number of pages, leaves, or columns in terms of the numbered or lettered sequences in the volume. Record the last numbered page, leaf, or column in each sequence[11] and follow it with the appropriate term or abbreviation.

>1 atlas (164 p.)
>
>1 atlas (24 leaves)
>
>1 atlas (xvi, 312 p., 212 p.)
>
>1 atlas (44 leaves, 50 p.)
>
>1 atlas (xii, a–h, 100, 32 p.)

Record pages, etc., that are lettered as inclusive lettering in the form *A–K p.*, *a–d leaves,* etc. Designate pages, etc., that are numbered in words or characters other than arabic or roman in arabic figures. (2.5B2)

>1 atlas (H–48, U–64, 32 p.)

5B6 Disregard unnumbered sequences, unless such a sequence constitutes the whole (see 5B10) or a substantial part (see also 5B11) of the publication, or unless an unnumbered sequence includes page(s), etc., that are referred to in a note. When recording the number of unnumbered pages, etc., either give the estimated number preceded by *ca.,* without square brackets, or enclose the exact number in square brackets.

>1 atlas (8, vii, ca. 300, 73 p.)
>
>1 atlas ([3], 10 leaves)
>
>1 atlas ([8], 155 p.)
>*Note:* Bibliography: 6th prelim. page

Disregard unnumbered sequences of inessential matter (advertising, blank pages, etc.). (2.5B3)

5B7 If the number printed on the last page or leaf of a sequence does not represent the total number of pages or leaves in that sequence, let it stand uncorrected unless it gives a completely false impression of the extent of the item, as, for instance, when only alternate pages are numbered or when the number on the last page or leaf of the sequence is misprinted. Supply corrections in such cases in square brackets.

(2.5B4)

>1 atlas (v, 296 [i.e. 122] p.)
>
>1 atlas (50 [i.e. 66] leaves)

5B8 If the numbering within a sequence changes (e.g., from roman to arabic numerals), ignore the numbering of the first part of the sequence. (2.5B5)

---

11. A sequence of pages or leaves is (1) a separately numbered group of pages, etc., (2) an unnumbered group of pages, etc., that stands apart from the other groups in the publication, or (3) a number of pages or leaves of plates distributed throughout the publication.

1 atlas (176 p.)
(Publication numbered i–xii, 13–176)

5B9   If the pages, etc., are numbered as part of a larger sequence (e.g., one volume of a multivolume publication) or if the item appears to be incomplete (see also 5B19), give the first and last numbers of the pages or leaves, preceded by the appropriate term or abbreviation.   (2.5B6)

1 atlas (leaves 81–149)

1 atlas (p. 713–797)

5B10   If the volume is printed without pagination or foliation, ascertain the total number of pages, etc., and give the number in square brackets. For larger items, estimate the number of pages and give the estimated number preceded by *ca.*, without square brackets.   (2.5B7)

1 atlas ([16] p.)

1 atlas ([93] p.)

1 atlas ([50] leaves)

1 atlas (ca. 600 p.)

5B11   If the volume has complicated or irregular paging, record the pagination using one of the following methods, depending upon the nature or extent of the complications:

a) Record the total number of pages or leaves (excluding those which are blank or contain advertising or other inessential matter) followed by the words *in various pagings* or *in various foliations*.

1 atlas (143 p. in various pagings)

1 atlas (256 leaves in various foliations)

b) Record the number of pages or leaves in the main sequences of the pagination and add the total number, in square brackets, of the remaining variously paged sequences.

1 atlas (226, [44] p.)

1 atlas (366, 98, [99] p.)

c) Describe the volume simply as *1 v. (various pagings), 1 case,* or *1 portfolio,* as appropriate (see 5B21).   (2.5B8)

1 atlas (1 portfolio (96 p. in various pagings))

5B12   Describe all loose-leaf publications that are designed to receive additions as *1 v. (loose-leaf), 2 v. (loose-leaf),* etc.   (2.5B9)

1 atlas (1 v. (loose-leaf))

5B13 Leaves or pages of plates. Record the number of leaves or pages of plates at the end of the sequence(s) of pagination, whether the plates are found together or distributed throughout the publication, or even if there is only one plate. For volumes consisting of unnumbered leaves or pages of plates, follow the instructions in 5B10. If the numbering of the leaves or pages of plates is complex or irregular, follow the instructions in 5B11.

> 1 atlas (246 p., 32 p. of plates)
>
> 1 atlas (xvi, 249 p., [12] leaves of plates)
>
> 1 atlas (x, 32, 73 p., [1] leaf of plates)
>
> 1 atlas ([16] p., [40] leaves of plates)
>
> 1 atlas ([80] p. of plates)
>
> 1 atlas (xii, 24 p., 212, [43] leaves of plates)

If the volume contains both leaves and pages of plates, record the number in terms of whichever is predominant. (2.5B10)

> 1 atlas (323 p., [19] p. of plates)
> *(Contains 16 pages and 3 leaves of plates)*

5B14 Describe folded leaves as such. (2.5B11)

> 1 atlas (44 folded leaves)
>
> 1 atlas (230 p., 25 leaves of plates (some folded))
>
> 1 atlas (25 folded leaves of plates)
>
> 1 atlas (1 portfolio (90 folded leaves of plates))

■APPLICATION Describe the leaves as found in the copy in hand, regardless of the state of other copies.

5B15 If numbered pages or leaves are printed on a double leaf (e.g., books in the traditional oriental format), record them as pages or leaves according to their numbering. If they are unnumbered, count each double leaf as two pages. (2.5B12)

> 1 atlas (26 leaves)
> *Note:* On double leaves, oriental style

5B16 If the paging is duplicated, as is sometimes the case with items having parallel texts, record both pagings and make an explanatory note (see 7B10).

(2.5B13 mod.)

> 1 atlas (xii, 35, 35 p.)
> *Note:* Opposite pages bear duplicate numbering

5B17 If a volume has a pagination of its own and also bears the pagination of a larger work of which it is a part, record the paging of the individual volume in this area and record the continuous paging in a note (see 7B10).   (2.5B14)

> 1 atlas (328 p.)
> *Note:* Pages also numbered 501–828

5B18   If the volume has groups of pages numbered in opposite directions, as is sometimes the case with items having texts in two languages, record the pagings of the various sections in order, starting from the title page selected for cataloguing.
                                                                                    (2.5B15 mod.)

> 1 atlas (ix, 155, 127, x p.)

5B19   If the last part of a publication is missing and the paging of a complete copy cannot be ascertained, give the paging in the form *234 + p.,* and make a note of the imperfection (see 7B20).   (2.5B16)

> 1 atlas (xxiv, 179 + p.)
> *Note:* Library's copy imperfect: all after p. 179 wanting

Publications in more than one volume

5B20   If an atlas is in more than one physical volume, record the number of volumes.
                                                                                    (2.5B17 mod.)

> 1 atlas (3 v.)

■APPLICATION   In describing a multivolume atlas that is not yet complete, precede the abbreviation for volumes by three spaces.

> 1 atlas (v.)

5B21   If the term *volume* is not appropriate for a multipart item, use one of the following terms:

Parts. Use for bibliographic units intended to be bound several to a volume, especially if so designated by the publisher.

Pamphlets. Use for collections of pamphlets bound together or assembled in a portfolio for cataloguing as a collection.

Pieces. Use for items of varying character (pamphlets, broadsides, clippings, maps, etc.) published, or assembled for cataloguing, as a collection.

Case(s). Use for either boxes containing bound or unbound material or containers of fascicles.

> 1 atlas (1 case (300 leaves of plates))

Portfolio(s). Use for containers holding loose papers, illustrative materials, etc. A portfolio usually consists of two covers joined together at the back and tied at the front, top, and/or bottom   (2.5B18)

> 1 atlas (12 portfolios)

5B22 If the number of bibliographic volumes differs from the number of physical volumes, record the fact in the form ₍bibliographic₎ v. in ₍physical₎.   (2.5B19)

>   1 atlas (8 v. in 5)

5B23 If a set of volumes is continuously paged, give the pagination in parentheses after the number of volumes. Ignore separately paged sequences of preliminary matter in volumes other than the first.   (2.5B20)

>   1 atlas (2 v. (xxxxi, 999 p.))

>   1 atlas (3 v. (xx, 800 p.))
>   *(Pages numbered xx, 1–201; xx, 202–513; xxi, 514–800)*

5B24 *Optional addition.* If the volumes in a multivolume set are individually paged, give the pagination of each volume in parentheses after the number of volumes.
           (2.5B21)

>   1 atlas (2 v. (xvi, 329; xx, 412 p.))

□ POLICIES

>   British Library
>   The option is applied.

>   Library of Congress
>   The option is applied selectively, e.g., continuously paged sets and early printed books.

>   National Library of Australia
>   The option is applied selectively.

>   National Library of New Zealand
>   The option is applied.

>   National Map Collection, PAC
>   The option is applied selectively, e.g., for early printed atlases.

5B25 If a publication was planned to be in more than one volume, but not all have been published and it appears that publication will not be continued, describe the incomplete set as appropriate (i.e., give paging for a single volume or number of volumes for multiple volumes), and make a note (see 7B10) to the effect that no more volumes have been published.   (2.5B22)

5B26 Braille or other raised types. If an item consists of leaves of braille or another tactile writing system, add, to the statement of the number of volumes or leaves, *of braille, of Moon type, of tactile maps, of tactile graphics,* etc., as appropriate.
           (2.5B23 mod.)

>   1 atlas (₍18₎ leaves of braille and tactile graphics)

>   1 atlas (51 leaves of braille and tactile graphics (14 folded))

1 atlas (310 leaves of braille)

1 atlas (200 leaves of tactile graphics)

Use *of jumbo braille* or *of microbraille* when appropriate.

1 atlas (4 v. of jumbo braille)

If an item contains press braille pages, add *of press braille* to the statement of the number of pages or volumes.

1 atlas (300 p. of press braille)

1 atlas (5 v. of press braille)

If an item consists of eye-readable print and braille, add *of print/braille* or *of print/ press braille,* as appropriate, to the statement of the number of pages or volumes.

(2.5B23)

1 atlas (300 p. of print/braille)

5B27   If the description is of a separately titled part of a cartographic item lacking a collective title (see 1G6), give the statement of extent as instructed in 5B1–5B3.

(3.5B4 mod.)

5B28   In describing a multipart item that is not yet complete, give the specific material designation alone preceded by at least three spaces.   (1.5B5 mod.)

₍*4 spaces*₎ maps

1 atlas ₍*3 spaces*₎ v.

Optionally, when the item is complete, add the number of physical units.   (1.5B5)

☐ POLICIES

British Library
The option is applied.

Library of Congress
The option is applied.

National Library of Australia
The option is not applied.

National Library of New Zealand
The option is applied.

National Map Collection, PAC
The option is applied.

Manuscript atlases

5B29   Single manuscripts. Record sequences of leaves or pages, whether numbered or not, as instructed in 5B4–5B25.

5B30 *Physical Description Area*

> (23 leaves)
>
> (iv, 103 leaves)
>
> ([63] leaves)
>
> ([4] p.)
>
> ([4], 103 p.)
>
> (leaves 51–71)

If the manuscript has been bound, add *bound* at the end of the pagination.

> ([70] leaves, bound)
>
> (4, [20], 30 p., bound)

Add, to the pagination, etc., of ancient, medieval, and Renaissance manuscripts, the number of columns (if more than one) and the average number of lines to the page.

> ([208] leaves (41 lines))
>
> ([26] leaves (2 columns, 45–47 lines))

Optional addition. Add, to the pagination, the number of leaves if this is different from the number of pages. (4.5B1)

> ([2] p. on 1 leaf)
>
> ([5] p. on 3 leaves)

□ POLICIES

> British Library
> The option is not applied.
>
> Library of Congress
> The option is not applied.
>
> National Library of Australia
> The option is not applied.
>
> National Library of New Zealand
> The option is not applied.
>
> National Map Collection, PAC
> The option is not applied.

Microforms

5B30 Record the number of physical units of a microform item by giving the number of parts in arabic numerals and one of the following terms as appropriate:

aperture card
microfiche
microfilm
microopaque

Optionally, if the general material designation *microform* is used, drop the prefix *micro* from these terms.

☐POLICIES
Library of Congress
The option is not applied.

National Library of Australia
The option is not applied.

National Library of New Zealand
The option is not applied.

National Map Collection, PAC
The option is not applied.

Add to *microfilm* one of the terms *cartridge, cassette,* or *reel*, as appropriate. Add to *microfiche* the term *cassette* if appropriate.

25 aperture cards

1 microfilm cassette

2 microfilm reels

3 microfiches

10 microopaques

Add the number of frames of a microfiche if it can be easily ascertained. Make the addition in parentheses.   (11.5B1)

1 microfiche (120 fr.)

5C   Other physical details   (3.5C)

5C1   Give the following details, as appropriate, in the order set out here:

number of maps in an atlas
colour
material
mounting   (3.5C1)

1 map : photocopy

2 maps on 1 sheet : both sides, col.

2 maps : back to back, col.

>   1 atlas (32 leaves) : ms., 24 col. maps, vellum
>
>   1 atlas (123 p.) : 35 col. maps (some folded)
>
>   1 model : col., plastic
>
>   1 globe : col., plastic, mounted on metal stand
>
>   1 view : ms., col., silk
>
>   1 map : ms., mounted on linen

■APPLICATION   If the method of production or reproduction is manuscript or photo-copy, record it as "other physical details." Record specific details on the method of production or reproduction in a note, if considered significant (7B10).

The revised citation order is:

>   layout on recto and verso, i.e., *back-to-back,* or *both sides*
>   (for maps only)
>   manuscript
>   photocopy
>   number of maps in an atlas
>   colour
>   material
>   mounting

For use of the terms *back-to-back* and *both sides,* see 5D1j Application.

5C2a   Specify the number of maps in an atlas as follows: (3.5C2)

>   1 atlas (xvi, 97, 100 p.) : 35 col. maps
>
>   1 atlas (330 p.) : 100 col. maps (some folded)
>
>   1 atlas ([13], 55, [15] p.) : 55 maps (chiefly col.)
>
>   1 atlas (1 portfolio ([15] folded leaves of plates)) : 15 folded col. maps
>
>   1 atlas ([13] leaves of braille and tactile graphics) : 6 col. tactile maps

5C2b   Include in a description of a printed atlas with illustrations the abbreviation *ill.* unless the illustrations are all of one or more of the particular types mentioned in the next paragraph. If only some of the illustrations belong to these types, give the abbreviation *ill.* first. Tables are not illustrations. Disregard illustrated title pages and minor illustrations (decorations, vignettes, etc.)   (2.5C1 mod.)

>   1 atlas (xix, 59 p.) : ill., 24 maps
>
>   1 atlas (ca. 350 p. (1 folded leaf of plates)) : ill. (some col.), 147 col.
> maps (1 folded)

5C2c   If the illustrations are of one or more of the following types, and are considered to be important, designate them by the appropriate term or abbreviation (in this order):

charts, coats of arms, facsimiles, forms, genealogical tables, maps, plans, portraits (use for both single and group portraits), samples. Designate all other types as *ill.*

(2.5C2 mod.)

1 atlas (lv, 48 p.) : ill., 45 maps, 39 ports.

1 atlas (viii, 236 p.) : ill., col. maps

1 atlas (xxxix, 352 p.) : ill., col. maps

5C2d Specify the number of illustrations if their number can be easily ascertained (e.g., when the illustrations are listed and their numbers stated). (2.5C4)

1 atlas (lv, 48 p.) : ill., 45 maps, 39 ports.

1 atlas (144 p., 300 p.) : ill., col. maps, 7 ports.

5C2e If some or all of the illustrations appear on the lining papers, make a note of this fact (see 7B10). (2.5C5)

1 atlas (viii, 236 p.) : ill., col. maps
Note: Maps on lining papers

5C2f Include illustrative matter issued in a pocket inside the cover of an item in the physical description. Specify the number of items and their location in a note (see 7B10). (2.5C7)

1 atlas (xiii, 417 p.) : ill., col. maps
Note: Six acetate overlays in separate envelope

1 atlas (xxxix, 352 p.) : col. ill., col. maps
Note: Two acetate maps, 3 coloured time charts (some folded) in pocket

1 atlas (v, 123 p.) : ill., maps
Note: Nine folded col. maps in pocket

1 atlas (2 v.) : col. maps (some folded)
Note: Two folded col. maps in jacket pocket (21 × 9 cm)

5C2g If rules 5C2a–5C2f give a misleading impression of the relative extent of the illustrations and text, combine the statements of pagination and of illustrative matter. (2.5C8)

1 atlas ([102] p. of various pagings, [58] leaves of maps)

5C3 Colour. If the item is coloured or partly coloured, indicate this. Disregard coloured matter outside a map, etc., border. (3.5C3)

180 maps : some col.

1 map : col.

4 maps : 2 col.

1 globe : col.

> 1 map : ms., col.

> 1 map : hand col.
> *(A printed map)*

Describe coloured illustrations in an atlas (i.e., those in two or more colours) as such.
(2.5C3 mod.)

> 1 atlas (xii, 24 p., xii) : col. ill., 30 col. maps

> 1 atlas (viii, 236 p.) : col. ill., col. maps

> 1 atlas (₍102₎ p. of various pagings, ₍58₎ leaves of plates) : ill., 58 hand col. maps

> 1 atlas (₍72₎ leaves, bound) : ms., col.

> 1 atlas (103 p.) : ill., hand col. maps, ports.

For slides and transparencies give an indication of the colour (col., b&w, etc.)
(8.5C12 mod., 8.5C16 mod.)

> 12 slides : sd. (3M Talking Slide), col.

> 3 transparencies (5 overlays each) : col.

■APPLICATION   If a printed item is hand coloured, state this as "hand col." Do not differentiate between colour and partial colour. Details on the amount of colour, or number of colours used, etc., may be recorded in a note (7B10). Monochromatic items are not considered coloured regardless of the colour of ink or paper used. If the colour of ink used on monochromatic items is other than black, or if the paper is other than white, this may be recorded in a note.

5C4   Material. Record the material of which the item is made if it is considered to be significant (e.g., if a map is on a substance other than paper).   (3.5C4 mod.)

> 1 map : col., plastic

> 1 map : col., silk

> 1 globe : col., wood

> 1 map : ms., col., vellum

5C5   Mounting. If the item is mounted or has been mounted subsequent to its publication, indicate this.   (3.5C5)

> 1 map : col., mounted on linen
> *Note:* Originally published in 4 sheets

> 1 globe : col., wood, mounted on brass stand

> 1 globe : plastic, mounted on metal stand

> 1 globe : col., relief pressed from metal

> 1 model : col., plastic

5C6 If a microform is negative, indicate this. (11.5C1)

> 1 microfilm reel : negative

5D Dimensions (3.5D)

■APPLICATION   In addition to the dimensions prescribed in 5D1– 5D8, the sheet size may be added whenever desired by the cataloguing agency. When recording more than one dimension separate each set with a comma.

> 1 map : col. ; 200 × 350 cm, folded to 20 × 15 cm, in plastic case 25
> × 20 cm

5D1a   Maps, plans, etc. For two-dimensional cartographic items, give the height × width in centimetres, to the next whole centimetre up (e.g., if a measurement is 37.1 centimetres, record it as 38 cm); *optionally,* for early and manuscript cartographic items, give the dimensions to the nearest millimetre. (See Figures 19– 33.) (3.5D1)

> 1 map : col. ; 25 × 35 cm
>
> 1 map : ms. ; 123.5 × 152.4 cm

■APPLICATION   When the map is placed in reading position, "height" is the top-bottom measurement of the map; "width" is the left-right measurement of the map.

Where the dimensions are given in centimetres and tenths of centimetres, always give the first significant number after the decimal, i.e., 126.0 × 340.0 cm means that the item has been measured in centimetres and tenths of centimetres and this item is exactly 126 centimetres high by 340 centimetres wide. When the dimensions are given in centimetres, any fraction being counted as a whole centimetre, express them in whole centimetres only, i.e., 126 × 340 cm means any fraction of a centimetre, if present, has been rounded up to the next whole centimetre.

As a matter of style and readability, do not repeat the word "on" in a statement of sheet size in the dimension element when the phrase "on sheet" is used in the extent of item element. (See Figure 26.)

□POLICIES
>   British Library
>   The option is applied.
>
>   Library of Congress
>   The option is not applied.
>
>   National Library of Australia
>   The option is applied.
>
>   National Library of New Zealand
>   The option is not applied.
>
>   National Map Collection, PAC
>   The option is applied.

5D1b Give the measurements of the face of the map, etc., measured between the neat lines. (3.5D1)

■APPLICATION The neat line is a line which encloses the detail of a map. There is only one neat line on a map. Measurement of a map includes any detail that breaks through the neat line. If there is no neat line, measure the maximum extent of the cartographic detail. (See Figures 19 and 20.)

For component maps of the same size, record the common size. If there are only two component maps or two common sizes, both may be given; otherwise record the greatest dimensions followed by ''or smaller.'' Always give the sheet size. (See Figure 21.)

5D1c Give the diameter of a circular map, etc., and specify it as such. (See Figures 22 and 23.) (3.5D1)

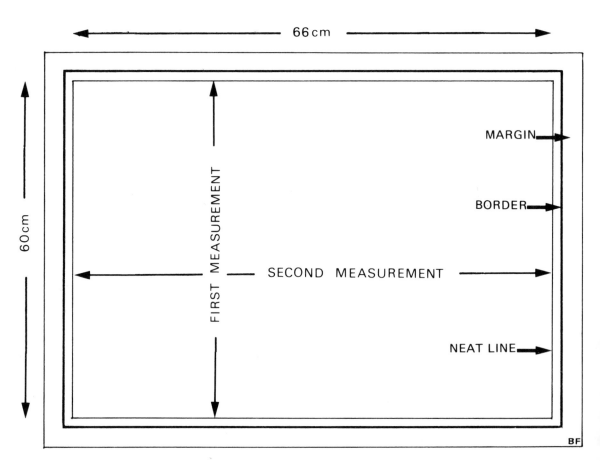

Figure 19 5D1b Dimensions. Neat line, border, and margins. Measure between neat lines

1 map : col. ; 60 × 66 cm

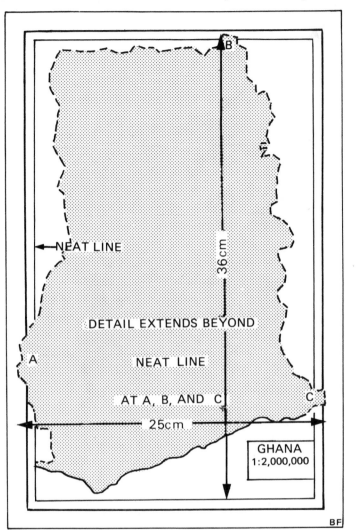

Figure 20 **5D1b** Dimensions. Detail extends beyond neat line

1 map : col. ; 36 × 25 cm

5D1d If a map, etc., is irregularly shaped, or if it has no neat line, or if it has bleeding or damaged edges, give the greater or greatest dimensions of the map itself. (See Figure 24.) (3.5D1 mod.)

> 1 map ; 78 × 80 cm
> *(Irregularly shaped)*

■APPLICATION The greater or greatest dimensions of the map itself are the extent of the cartographic content, coverage, or image on the item. Give these dimensions if it is

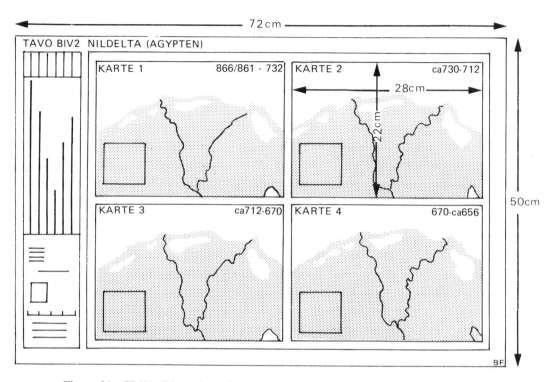

Figure 21   5D1b   Dimensions. Component maps

4 maps on 1 sheet : col. ; each 22 × 28 cm, sheet 50 × 72 cm

possible to measure the height and width of the map itself using a maximum of two point to point measurements.

5D1e   If it is difficult to determine the points for measuring the height and the width of the map, etc. (e.g., when the shape is extremely irregular, or when it was printed without one or more of its borders, or when it lacks one or more of its borders), give the height × width of the sheet specified as such.   (3.5D1)

> 1 map : col. ; on sheet 45 × 33 cm

5D1f   Measure a single map, etc., drawn in segments at a consistent scale as if it were joined. Add the sheet size. If such a map, etc., has been mounted, give the dimensions of the whole map, etc., alone. (See Figures 25, 26, and 27.)   (3.5D1 mod.)

> 1 map ; 10 × 60 cm, on sheet 25 × 35 cm
>
> 1 map on 9 sheets ; 264 × 375 cm, sheets 96 × 142 cm
>
> 1 map on 4 sheets ; sheets 30 × 40 cm
>
> 1 map on 2 sheets : ms., col. ; 47 × 229 cm, sheets 49 × 119 cm
> *Note:* Map restored. Original folded dimensions, 21 × 15 cm

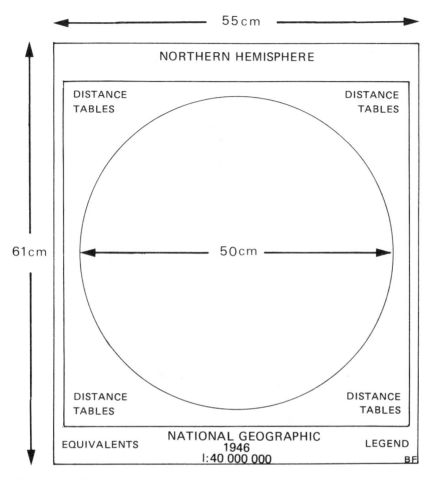

Figure 22 5D1c Dimensions. Circular map

1 map : col. ; 50 cm diam., on sheet 61 × 55 cm

■APPLICATION   If the sheets are numerous (i.e., more than 15), and the assembly of them results in an irregularly shaped map, and if measuring that assembled map is very difficult, then record the dimensions of the largest sheet followed by the words *or smaller* if appropriate.

This rule does not apply to series maps.

5D1g   If the size of either dimension of a map, etc., is less than half the same dimension of the sheet on which it is printed or if there is substantial additional informa-tion on the sheet (e.g., text), give the sheet size as well as the size of the map, etc. (See Figure 28.)   (3.5D1)

1 map ; 20 × 31 cm, on sheet 42 × 50 cm

■APPLICATION   Substantial additional information on the map sheet could be text,

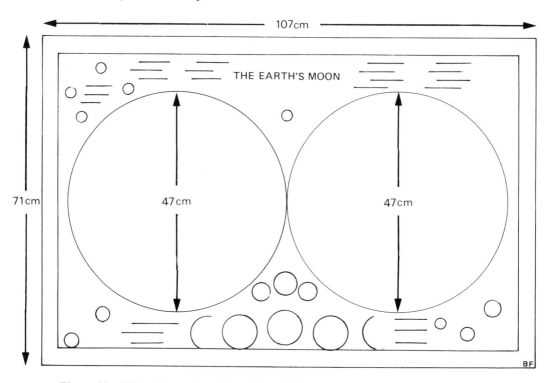

Figure 23 5D1c Dimensions. Double hemisphere

1 map : col. ; 2 hemispheres each 47 cm in diam., on sheet 71 × 107 cm

legend, photographs, insets, or ancillary maps, in which case always give the sheet size. (See Figure 28.)

5D1h If a map, etc., is printed with an outer cover within which it is intended to be folded or if the sheet itself contains a panel or section designed to appear on the outside when the sheet is folded, give the sheet size in folded form as well as the size of the map, etc. (See Figure 29.) (3.5D1)

> 1 map ; 80 × 57 cm, folded to 21 × 10 cm
>
> 1 map : col. ; 9 × 20 cm, on sheet 40 × 60 cm, folded to 21 × 10 cm

■APPLICATION This rule refers to items which are designed to be folded but are available flat or folded. Base the measurements on the size of the panel, section, etc. This rule does not apply to items which may have been folded for convenience (e.g., by publisher, manufacturer, distributor, or subsequent owners) but have not been designed to be folded.

5D1j If a map, etc., is printed on both sides of a sheet at a consistent scale, give the dimensions of the map, etc., as a whole, and give the sheet size. If such a map, etc., cannot conveniently be measured, give the sheet size alone. (See Figure 30.) (3.5D1)

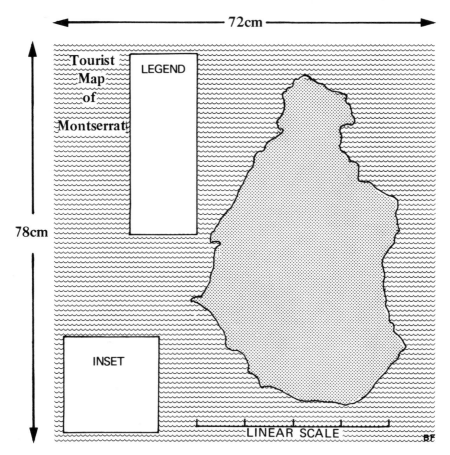

Figure 24 5D1d Dimensions. Bleeding edge

1 map : col. ; 78 × 72 cm

 1 map : both sides ; 45 × 80 cm, on sheet 50 × 44 cm
   (Printed on both sides of sheet with line for joining indicated)

 1 map : both sides ; on sheet 45 × 30 cm
   (Printed on both sides of sheet)

■APPLICATION   Every effort should be made to record the size of the map, eliminating any overlap in coverage.

   If an inset is a continuation of the main map, treat it in the note area as an inset, even if it is at the same scale. (See Figure 31.)

   If a cartographic item has maps printed on both sides and a collective title (or supplied collective title) and if the collective title is used in the description, record the dimensions of the component maps and the sheet as instructed in 5D1b Application.

   5 maps on 1 sheet : both sides, col. ; 70 × 50 cm or smaller, sheet 80 × 102 cm

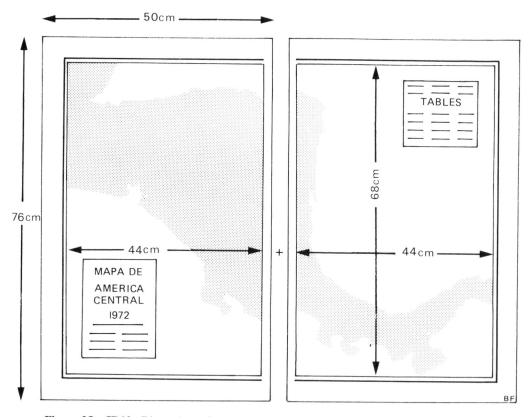

Figure 25   5D1f   Dimensions. One map on two or more sheets

1 map on 2 sheets : col. ; 68 × 88 cm, sheets 76 × 50 cm

If a cartographic item has maps printed on both sides, one of which is a translation of the other, record the dimensions of the map. Add the sheet size if considered necessary.

2 maps on 1 sheet : back to back, col. ; 60 × 80 cm, sheet 65 × 80 cm

5D1k   If the maps, etc., in a collection are of two sizes, give both. If they are of more than two sizes, give the greatest height of any of them followed by greatest width of any of them and the words *or smaller*.   (3.5D1)

60 maps ; 44 × 55 cm and 48 × 75 cm

60 maps ; 60 × 90 cm or smaller

10 maps in 25 sheets : col. ; sheets 100 × 90 cm or smaller

■APPLICATION   For series maps, or collections of maps, record the map size and, if appropriate, the sheet size. If there are two sizes, record both. (See Figure 32.) Such maps may be printed on standard size sheets, but they may not all be printed in the same direction on the sheet. For example, the sheets in Figure 33 are the same size, but when put in a reading position the entry varies from that of a single sheet.

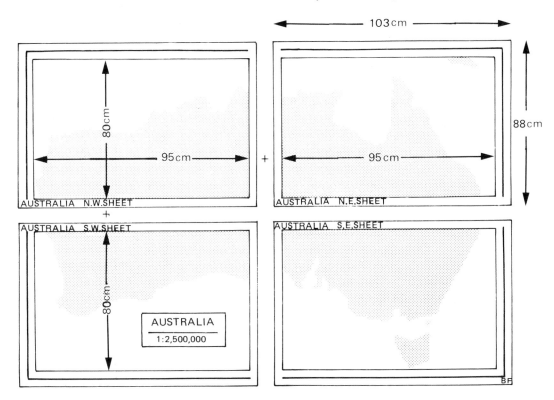

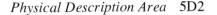

Figure 26   5D1f   Dimensions. One map on two or more sheets

1 map on 4 sheets : col. ; 160 × 190 cm, sheets 88 × 103 cm

If there are more than two sizes, give the greatest height of any of them followed by the greatest width of any of them and the words *or smaller* (see Figure 34).

This rule can also be applied to an incomplete map published in more than one sheet for which it is difficult to establish the dimensions of the whole map. In this case give the greatest height and the greatest width of the sheets held and add the phrase *or smaller*.

> ; sheets 66 × 108 cm or smaller

5D2   Atlases. Give the height of the volume(s) in centimetres, to the next whole centimetre up (e.g., if a volume measures 17.2 centimetres, record it as 18 cm). Measure the height of the binding if the volume is bound. Otherwise, measure the height of the item itself. If the volume measures less than 10 centimetres, give the height in millimetres.   (2.5D1, 3.5D2)

> 1 atlas (xii, 100, 32 p.) : 100 col. maps ; 29 cm

> 1 atlas (3, [3], 95 [i.e. 97] p.) ; 37 maps ; 60 cm
> *Note:* Rebound. Original size ca. 46 cm

If the width of the volume is either less than half the height or greater than the height, give the width following the height preceded by a multiplication sign.   (2.5D2)

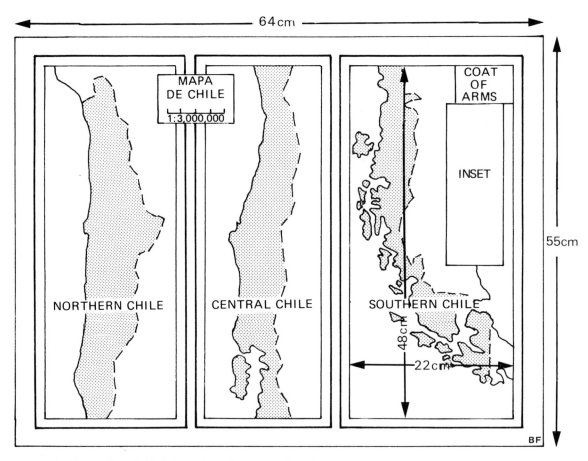

Figure 27 5D1f Dimensions. One map printed in segments on one sheet

1 map : col. ; 144 × 22 cm, on sheet 55 × 64 cm
(Explain the layout in a note, if necessary)

; 20 × 8 cm

; 20 × 32 cm

If the volumes in a multivolume set differ in height and the difference is two centimetres or less, give the largest size. If the difference is more than two centimetres, give
the smallest size and the largest size, separated by a hyphen.   (2.5D3)

; 24– 28 cm

If the volume consists of items of varying height bound together, give the height of
the binding only.   (2.5D5)

5D3 Relief models. For relief models, give the height × width in centimetres as
instructed in 5D1, and *optionally* add the depth.   (3.5D3)

116

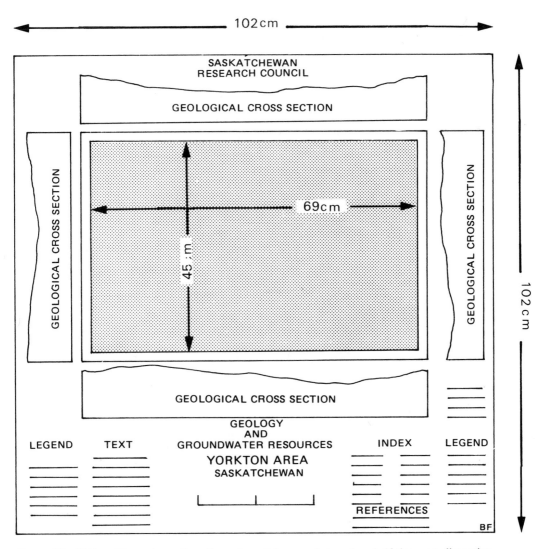

Figure 28 5D1g Dimensions. One dimension of the map is less than half the same dimension of the sheet

1 map : col. ; 45 × 69 cm, on sheet 102 × 102 cm

1 model : col., plastic ; 45 × 35 × 2 cm

☐ POLICIES
British Library
The option is applied.

Library of Congress
The option is applied.

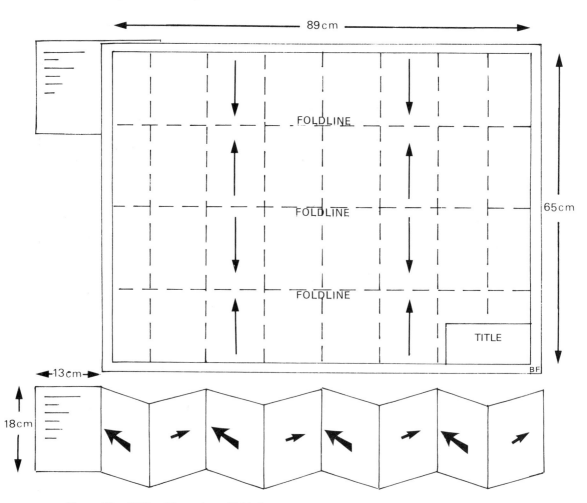

Figure 29   5D1h   Dimensions. Folded map

1 map : col. ; 65 × 89 cm, folded to 18 × 13 cm

National Library of Australia
  The option is applied.

National Library of New Zealand
  The option is applied.

National Map Collection, PAC
  The option is applied.

5D4   Globes. For globes, give the diameter, specified as such.   (3.5D4)

   1 globe : col., wood, mounted on metal stand ; 12 cm in diam.

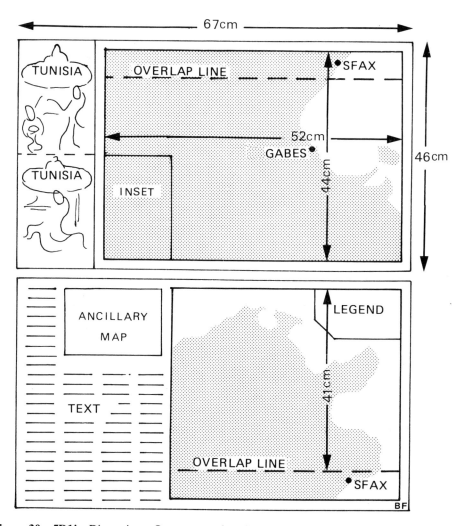

Figure 30 5D1j Dimensions. One map continued on verso

1 map : both sides, col. ; 85 × 52 cm, on sheet 46 × 67 cm

5D5 *Optional addition.* **Containers.** Add the dimensions of a container, specified as such, to the dimensions of the item. (3.5D5)

    1 globe : col., plastic, mounted on metal stand ; 20 cm in diam., in box 40 × 21 × 21 cm

    1 map : col. ; 200 × 350 cm, folded to 20 × 15 cm, in plastic case 25 × 20 cm

    1 map : col. ; 48 × 49 cm, folded in envelope 23 × 31 cm

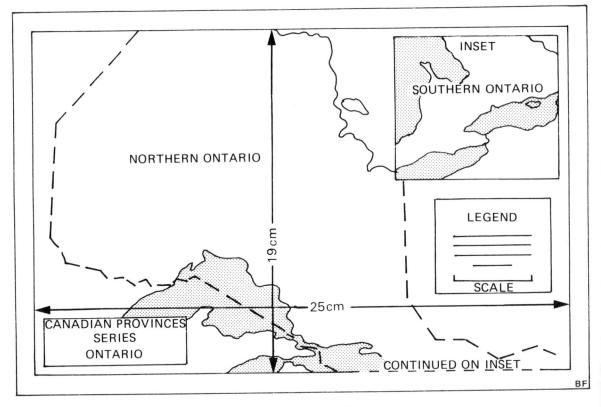

Figure 31   5D1j   Dimensions. One map continuing in an inset

1 map ; 19 × 25 cm

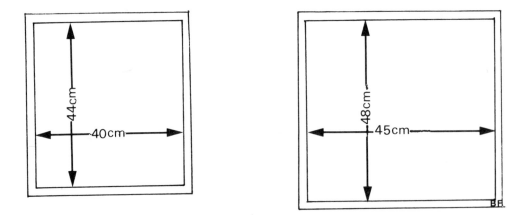

Figure 32   5D1k   Dimensions. Two sizes of maps in a series

; 44 × 40 cm and 48 × 45 cm

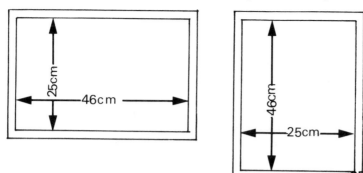

Figure 33   5D1k   Dimensions. Two sizes of maps in a series
; 25 × 46 cm and 46 × 25 cm

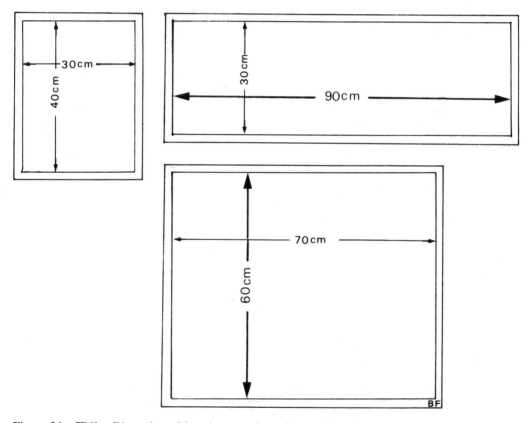

Figure 34   5D1k   Dimensions. More than two sizes of maps in a series
; 60 × 90 cm or smaller

☐ POLICIES

British Library
The option is applied.

Library of Congress
The option is applied.

National Library of Australia
The option is not applied.

National Library of New Zealand
The option is applied.

National Map Collection, PAC
The option is applied on a case by case basis.

5D6   For a map on a transparency, or a map mounted in a frame, give the height and the width of the item, excluding any frame or mount. Details on mounting may be recorded in a note. (See 7B10.)   (8.5D4 mod.)

3 transparencies (15 overlays) : b&w ; 26 × 22 cm

5D7   For cartographic items on a slide give the height and the width only if the dimensions are other than 5 × 5 cm (2 × 2 in.).   (8.5D5 mod.)

1 slide : col.

1 slide : b&w ; 7 × 7 cm

5D8   Microfiches. Give the height × width of a microfiche in centimetres. Record a fraction of a centimetre as the next whole centimetre up.   (11.5D3 mod.)

3 microfiches ; 10 × 15 cm

5E   Accompanying material   (3.5E)

5E1a   There are four ways of recording information about accompanying material:

    i)   record the details of the accompanying material in a separate entry

or ii) record the details of the accompanying material in a multilevel description (see 13F)

or iii) record the details of the accompanying material in a note (see 7B11).

Accompanied by atlas "A demographic atlas of Northwest Ireland" (39 p. : col. maps ; 36 cm), previously published separately in 1956

Teacher's guide (24 p.) by Robert Garry Shirts

Accompanied by filmstrip entitled: Mexico and Central America

1 atlas (132 p.) : chiefly maps ; 24 cm
Note: Two col. transparencies (23 × 8 cm) inserted

or iv) record the name of accompanying material at the end of the physical description.

(1.5E1)

387 p. : ill. ; 27 cm + teacher's notes

32 p. : col. ill. ; 28 cm + 3 maps

271 p. : ill. ; 21 cm + 1 atlas

17 maps ; 90 × 96 cm + 1 book (xvii, 272 p. ; 25 cm)

1 atlas (16 leaves) : chiefly maps ; 30 cm + legends

maps : col. ; 76 × 56 cm, folded to 19 × 14 cm + booklets

1 atlas (₍44₎ leaves of plates) : maps (some col.) ; 43 cm + texts (2 v. :
ill. ; 30 cm)

1 atlas (ca. 300 p.) : chiefly col. maps ; 50 cm + index (1032 p. ; 25
cm)

1 atlas (₍60₎ leaves of plates) : maps ; 48 × 62 cm + text (36, ₍2₎ p., ₍1₎
leaf of plates : folded map ; 20 cm)

maps : col. ; 53 × 79 cm or smaller, folded in cover 63 × 49 cm +
legend (₍1₎ leaf, 12 p. ; 61 cm)

1 map : col. ; 66 × 55 cm, folded to 17 × 15 cm + text (168 p. ; 24
cm)

■APPLICATION   For maps include relatively simple, non-complicated accompanying
material with the physical description; but transcribe more complex bibliographic information for more significant accompanying material in a note (7B11), e.g., if the accompanying material has a distinctive title.

5E1b   *Optional addition.* If method iv is applicable and further physical description is
desired, add a statement of the extent, other physical details, and dimensions of the
accompanying material as appropriate. Formulate such additional descriptions in accordance with the rules for the material or type of publication to which the accompanying
material belongs.

(1.5E1)

. . . 21 cm + 1 atlas (301 p. : ill. (some col.) ; 23 cm)

. . . 18 cm + 20 slides (col.)

17 maps ; 90 × 96 cm + 1 book (xvii, 272 p. ; 25 cm)

6   SERIES AREA

(3.6)

Contents:

6A   Preliminary rule
6B   Title proper of series
6C   Parallel titles of series
6D   Other title information of series

6A Preliminary rule   (3.6A)

6A1 Punctuation

For instructions on the use of spaces before and after prescribed punctuation, see 0C.
Precede this area by a full stop, space, dash, space.
Enclose each series statement (see 6J) in parentheses.
Precede parallel titles of series or subseries by an equals sign.
Precede other title information relating to series or subseries by a colon.
Precede the first statement of responsibility relating to a series or subseries by a diagonal slash.
Precede subsequent statements of responsibility relating to a series or subseries by a semicolon.
Precede the ISSN of a series or subseries by a comma.
Precede the numbering within a series or subseries by a semicolon.
Precede the title of a subseries by a full stop.   (3.6A1)

. — (Title proper of series = parallel title of series / statement of responsibility relating to series ; numbering within series. Title of subseries = parallel title of subseries ; numbering within subseries)

. — (Title proper of series / statement of responsibility relating to series, ISSN of series ; numbering within series. Title of subseries / statement of responsibility relating to subseries, ISSN of subseries ; numbering within subseries)

. — (Title proper of series / statement of responsibility relating to series ; numbering within series)

. — (Title proper of series : other title information of series / statement of responsibility relating to series ; numbering within series)

. — (Title proper of series / statement of responsibility relating to series ; numbering within series) (Title proper of series / statement of responsibility relating to series ; numbering within series)

6A2 Sources of information.
Take information recorded in this area from the chief source of information or from any other source specified for this area in the following rules. Enclose any information supplied from other sources in square brackets, within the parentheses enclosing each series statement.   (1.6A2)

6B Title proper of series   (1.6B)

6B1 If an item is one of a series, record the title proper of the series as instructed in 1B (see also 6B2).   (1.6B1)

Climatological studies ; no. 8

A1 street atlas series

Carte géographique de l'Angleterre ; no 16

Deutscher Planungsatlas ; Bd. 8

Bartholomew world travel series

Nouvelle collection / Maurice Le Lannou

Communications of the Dublin Institute for Advanced Studies. Series D, Geophysical bulletin ; no. 29

Series of atlases in facsimile / Theatrum Orbis Terrarum. 6th series ; v. 1

Saggi e memorie di storia dell'arte ; v. 7

Graeco-Roman memoirs, ISSN 0306-9992 ; no. 93

Military city map : series A902 1:25 000 ; MCE 329 = Carte militaire de la ville : série A902 1:25 000 ; MCE 329

Série SIPE

Les Atlas Jeune Afrique

The heritage collection illustrated atlas from Unigraphic

Livrecartes

■APPLICATION   If the title proper of a series includes a series number, record that number as part of the title proper of the series.

Faeroe Islands 1:20,000, Geographical Section, General Staff, Series GSGS 4367

Nigeria 1:62,500 preliminary plot, Directorate of Colonial Surveys, Series DCS 32

Korea 1:25,000, Series L851

6B2   Cartographic serials (not series). In case of doubt about whether a corporate body's name or an abbreviation of that name is part of the title proper, treat the name as such only if it is consistently so presented in various locations in the serial (cover, caption, masthead, editorial pages, etc.) and/or, when cataloguing retrospectively, in indexes, abstracts, or other lists.   (12.1B2 mod.)

6B3   If variant forms of the title of the series (other than parallel titles) appear, choose the title given in the chief source of information as the title proper of the series. Give the variant form(s) in the note area if it is of value in identifying the item.

If variant forms of the title of the series (other than parallel titles) appear in the chief source of information, use the variant that identifies the series most adequately and succinctly.

If the title of the series does not appear in the chief source of information and variant forms appear elsewhere in the publication, choose the variant that identifies the series most adequately and succinctly, preferring a variant that appears in the other preliminaries. (1.6B2)

6C   Parallel titles of series (1.6C)

6C1   Follow the instruction in 1D (second level of description) when recording the parallel titles of a series. (1.6C1)

> Jeux visuels = Visual games

6D   Other title information of series (1.6D)

6D1   Include other title information of a series only if it provides valuable information identifying the series. Follow the instructions in 1E when recording other title information of a series. (1.6D1)

6E   Statements of responsibility relating to series (1.6E)

6E1   Give statements of responsibility appearing in conjunction with the series title if they are considered to be necessary for identification of the series. Follow the instructions in 1F when recording a statement of responsibility relating to a series. (1.6E1)

> Map supplement / Association of American Geographers
>
> Technical memorandum / Beach Erosion Board
>
> Research monographs / Institute of Economic Affairs
>
> Special publication / Colorado Geological Survey ; 5-B
>
> Map / Geological Survey of Alabama ; 137
>
> Map and chart series / Geological Society of America ; MC-27
>
> Dossiers cartographiques / Québec Ministère de l'industrie et du commerce ; 2
>
> Map abstract / Dept. of Geology and Geography, University of Alabama ; no. 5
>
> Resource map / New Mexico Bureau of Mines and Mineral Resources ; 10

6F   ISSN of series (1.6F)

6F1   Record the International Standard Serial Number (ISSN) of a series if it appears in the item being described (see also 6H4). Record the ISSN in the standard manner, i.e., ISSN followed by a space and two groups of four digits separated by a hyphen.

(1.6F1)

> Western Canada series report, ISSN 0317-3127

6G   Numbering within series   (1.6G)

6G1   Record the numbering of the item within the series in the terms given in the item. Use standard abbreviations (see Appendix H) and substitute arabic numerals for other numerals or spelled out numbers (see Appendix J).   (1.6G1)

> Suomalaisen Kirjallisuuden Seuran toimituksia, ISSN 0355-1768 ; 325

> Environment science research ; v. 6

■APPLICATION   If in a parallel series statement the numbering of the item within the series appears only once but the design of the chief source of information makes it clear that it is intended to be read more than once, repeat the series numbering without the use of square brackets. (See example in rule 6H3.)

6G2   If the parts of a multipart item are separately numbered within a series, give the first and the last numbers if the numbering is continuous; otherwise, give all the numbers.

(1.6G2)

> ; v. 11–15

> ; v. 131, 145, 152

For numbering of cartographic serials, record each series statement as instructed in 6. Do not give series numberings if each issue is separately numbered within the series.

(12.6B1 mod.)

6G3   If the item has a designation other than a number, give the designation as found.

(1.6G3)

> ; v. A

> ; 1971

6H   Subseries   (1.6H)

6H1   If an item is one of a subseries (a series within a series, whether or not it has a dependent title) and both the series and the subseries are named in the item, give the details of the main series (see 6A–6G) first and follow them with the name of the subseries and the details of that subseries.   (1.6H1)

> Geological Survey professional paper ; 683–D.   Contributions to
> palaeontology

6H2   If the subseries has an alphabetic or numeric designation and no title, give the designation. If such a subseries has a title as well as a designation, give the title after the designation.   (1.6H2)

>   Historischer Atlas von Bayern.   Teil Altbayern ; Heft 47

6H3   Add parallel titles, other title information, and statements of responsibility relating to subseries as instructed in 6C, 6D, and 6E.   (1.6H3)

>   Papers and documents of the I.C.I.   Series C, Bibliographies ; no. 8 =
>   Travaux et documents de l'I.C.I.   Série C, Bibliographies ; no. 8

6H4   Add the ISSN of a subseries if it appears in the item being described; in such a case, omit the ISSN of the main series.   (1.6H4)

>   Janua linguarum.   Series maior, ISSN 0075-3114

>   *not*   Janua linguarum, ISSN 0446-4796.   Series maior, ISSN 0075-3114

6H5   Add the numbering within a subseries as instructed in 6G.   (1.6H5)

6J   More than one series statement   (1.6J)

6J1   The information relating to one series or subseries constitutes collectively one series statement. If an item belongs to two or more series and/or series and subseries, make separate series statements and enclose each statement in parentheses. Follow the instructions in 6A–6H in recording each series statement. If the criterion applies, give the more specific series first.

>   (Centre de recherches d'histoire ancienne ; v. 35) (Annales littéraires de
>   l'Université de Besançon ; 240)
>   (*Title of item:* Carte archéologique du Cher . . .)

If parts of an item belong to different series and this relationship cannot be stated clearly in the series area, give details of the series in a note (see 7B12).

                                                                                    (1.6J1)

7   NOTE AREA

Contents:

7A	Preliminary rule
7A1	Punctuation
7A2	Sources of information
7A3	Form of notes
	Order of information
	Script
	Quotations
	References
	Formal notes
	Informal notes
7A4	Notes citing other editions and works
7A5	Scope

7A   Preliminary rule[12]   (3.7A)

7A1   Punctuation. Precede each note by a full stop, dash, space *or* start a new paragraph for each.

Separate introductory wording from the main content of a note by a colon and a space.   (3.7A1)

7A2   Sources of information. Take data recorded in notes from any suitable source. Use square brackets only for interpolations within quoted material. (See also 0E.)   (1.7A2)

---

12. The rules in 7A provide guidance for formulating the notes outlined in 7B. They are not part of the order to be followed in the note area which is given in 7B.

7A3   Form of notes

Order of information. If data in a note correspond to data found in the title and statement of responsibility, edition, material (or type of publication) specific details, publication, etc., physical description, and series areas, give the elements of the data in the order in which they appear in those areas. In such a case, use prescribed punctuation, except substitute a full stop for a full stop, space, dash, space.

> American ed.: New York Times atlas of the world.   New York : New York Times Book Co., 1978

> Translation of: Archaeological atlas of the world / David and Ruth Whitehouse

> Originally published: London : Gray, 1871

> Revision of: 3rd ed.   London : Macmillan, 1953

When giving in notes names or titles originally in nonroman scripts, use the original script whenever possible rather than a romanization.

> Based on: Братья Карамазовы / Ф. М. Достоевский

Quotations. Give quotations from the item or from other sources in quotation marks. Follow the quotation by an indication of its source, unless that source is the chief source of information. Do not use prescribed punctuation in quotations.

> "Published for the Royal Institute of Public Administration"

> "A textbook for 6th form students"—Pref.

> "Todas las cartas geográficas que forman el presente atlas, fueron dibujadas por el Prof. José Santiago Léon"

> "Photography and index by Mark Hurd Aerial Surveys, Inc."—On most maps

References. Make reference to passages in the item, or in other sources, if these either support the cataloguer's own assertions or save repetition in the catalogue entry of information readily available from other sources.

> Cf.: Catalogue of the Henry Newton Stevens collection of the Atlantic Neptune / H.N. Stevens

> "This seems to be the earliest pocket edition of Speed's book of letterpress and maps"—Hazlitt, Collections and notes, v. 1, p. 397

> "Most publications today take Cresques Abraham to be the author of the Paris Atlas . . . yet even this authorship is not certain"—P. 12

> "Knox, the bookseller, is said to be the real compiler"—Watt, Bibl. Brit., v. 1, p. 452
> > (*On t.p.*: A universal geography . . . originally compiled by William Guthrie.   Accompanying atlas: General atlas for Guthrie's geography)

Formal notes. Use formal notes employing an invariable introductory word or phrase or a standard form of words when uniformity of presentation assists in the recognition of the type of information being presented or when their use gives economy of space without loss of clarity.

Informal notes. When making informal notes, use statements that present the information as briefly as clarity, understandability, and good grammar permit.   (1.7A3)

7A4   Notes citing other editions and works

Other editions. In citing another edition of the same work, give enough information to identify the edition cited.

> Revision of: 2nd ed., 1973

> Simultaneously published: Toronto : Fitzhenry & Whiteside

Other works and other manifestations of the same work. In citing other works and other manifestations of the same work (other than different editions with the same title), always give the title and (when applicable) the statement(s) of responsibility. Give the citation in the form, author, title proper; *or* in the form, title proper / statement of responsibility. When necessary, add the edition and/or date of publication of the work cited.   (1.7A4)

> Translation of: Atlas rybacki szelfu Afryki polnocno-zachnodniej / Andrzej Klimaj. v. 1–1971, v. 2–1973

> Sequel to: Past and present of Allamakee County, Iowa / John Hancock

> Precursor of: Flora of the Great Plains

> Continues: General map of United States east, main automobile routes / American Automobile Association

> Supersedes: NAVAIR 50-1C-52. 1966

> Enlarged from: Mikrókosmos / Peter Heylyn. Oxford, 1621

> Abridgement of: Atlas Marianus / Wilhelm Gumppenberg. 1657–1659

7A5   Notes contain useful descriptive information that cannot be fitted into other areas of the description. A general outline of notes is given in 7B. When appropriate, combine two or more notes to make one note.   (1.7A5 mod.)

7B   Notes
Make notes as set out in the following subrules and in the order given there.[13]

                    (3.7B)

---

13. As the notes given from *AACR2* 3.7B1 to 3.7B21 are not sufficiently comprehensive for cataloguing cartographic materials, a number of additional notes have been appropriated from other chapters of *AACR2* and inserted in their correct sequence according to the general outline given in *AACR2* 1.7B. One new note is included (relief features), and has been incorporated in 7B1.

7B1a Nature and scope of the item. If the nature or scope of a cartographic item is not apparent from the rest of the description, indicate it in a word or brief phrase. Also give a note on unusual or unexpected features of the item. (3.7B1)

> Shows all of western Europe and some of eastern Europe
> (*Item entitled:* Germany)

> Shows the routes of Amundsen, Byrd, and Gould

> Shows southernmost extent of the midnight sun

> Shows dioceses

> A collection of photocopies of maps from various sources

> Also shows geologic formations
> (*Title proper:* Mineral resources map of Wyoming)

> Maps are photographs of col. relief models with "order of battle data, place names, etc." added

> Points of interest shown pictorially

> Does not show land capability
> (*Title proper:* Bennington County, Vermont, land capability plan)

> Covers area between Cape Cod and Assateague Island
> (*Title proper:* Bathymetric map, Atlantic coast, United States)

> Shows only the continental United States
> (*Title proper:* Portrait map of the United States of America)

■APPLICATION   Include information on subject content here.
  Include the frequency of publication of cartographic serials here.
  Situation date. The situation date (or dates) of the information shown on the map is recorded here, e.g., statistical information, road revision, historical events, cultural features, surveys, expeditions. It is always recorded if it differs from the publication or edition date. A feature date or an imaginary date may also be recorded. The notes should indicate the earliest and latest dates recorded; they should also indicate, if mentioned, to which specific features the dates refer.

> Shows the main battles of 1944–1945
> (*Item entitled:* The Asian struggle)

> Based on 1961 statistics

> Roads revised 1967

> Based on 1961 census

> Data for 1970–1975

> Date of situation: 1961–1970
> (*On item:* roads revised to 1969–70, cultural features 1961)

> Maps compiled in 1975 from 1970 census data

> Includes actual data for 1977, and preliminary data for 1978

The source of information for the situation date should be taken from a date printed on the map itself, from dates deduced from information shown on the map, from accompanying text, or from any other source. Indicate if the date is approximate or unreliable.

> Railways revised 1857

> Depicted as built, railways not yet constructed at that date

> "Data used cover a total of twenty years: 1951–1971"—Intro.

See also Appendix C.

Relief. Indicate the method of relief portrayal here. (See Figure 35.)

> Depths shown by contours and soundings

> "Contour interval 20 feet"

> Relief shown by contours, hachures, and spot heights

> Relief shown pictorially

Manuscripts

7B1b   Nature, scope, or form of manuscript(s). Make notes on the nature of a manuscript or the manuscripts in a collection unless it is apparent from the rest of the description. (4.7B1 mod.)

7B1c   If the item or collection being described is a copy or consists of copies, add *(carbon copy), (transcript),* or the plural of these as appropriate. To transcript, add *handwritten* or *typewritten*. (4.7B1 mod.)

7B1d   If the items in a collection are not all of the same nature, word the qualification to indicate this. (4.7B1)

> Some ms. (some photocopies)

> Some ms. (transcripts, and photocopies)

7B1e   If the manuscript item is signed, indicate this. (4.7B1)

> Signed ms.

7B1f   If the item is a copy, add the location of the original if this can be readily ascertained. (4.7B1)

> Original in the British Library Reference Division

7B1g   In describing a collection of manuscripts, name the types of papers, etc., constituting the collection and mention any other features that characterize it. If the collection is of personal papers, give enough data to identify the person, either as a brief initial statement or as part of the summary of the nature of the collection. If necessary, give the contents (see also 7B18) as part of that summary. (4.7B1)

Paleontologist and educator. Correspondence, reports, notes, articles, maps, printed matter, and other papers, mainly relating to the Carnegie Institution, the National Academy of Sciences, the National Research Council, and national parks

Papers covering (in the main) Allen's service as U.S. senator, 1837–1848, and as governor of Ohio, 1873–1874. Includes some of his speeches, drafts of his letters, and letters from various correspondents on political matters in Ohio

Includes records of the Banking Board, 1911–1939, and those of the Bureau of Insurance 1897–1943

Writer. Personal papers, letters, etc., drafts of some poems, including the complete text of the verse drama "The Pierrot of the minute"

7B2   Language. Give the language or languages of captions, etc., and text unless they are apparent from the rest of the description.   (3.7B2)

In Esperanto

Includes text in Finnish, Swedish, English, and German

Place names in Italian

Legend in Afrikaans and English

Except for title and "La mer du Nord" the map is in English

English and Japanese, with ms. Spanish translations of Japanese captions

Table of contents, abstract, and legend in English, Russian, and Spanish.   Added t.p. in English and Spanish

Explanatory text in English, with French summaries of each chapter

Arabic braille with brief English ms. captions

English and romanized Chinese (Pinyin)

Pinyin place names, with conversion table from Wade-Giles

■APPLICATION   This includes notes on translations (see *AACR2* 1.7B2).

7B3   Source of title proper. Make notes on the source of the title proper if it is other than the chief source of information. *Optionally,* make notes on the location of the title proper taken from the chief source of information.   (3.7B3 mod.)

Title from separate wrapper

Title from: A list of maps of America / P.L. Phillips.   p. 502

Title from container

Panel title

Title from label on case

Title from label (in ms.) pasted on cover

Title from portfolio

Title from Kümmerly & Frey Gesamtkatalog, 1948

Title from limited ed. statement on 79th leaf

Title from maps
   (For atlas)

Caption title
   (For atlas)

Cover title

Title from publisher's packing lists and correspondence

Title from broadside pasted on prelim. leaf

Title from typed sheet attached to first map

Title from photocopy of 1879 ed.

□ POLICIES
   British Library
      The option is applied on a case by case basis.

   Library of Congress
      The option is applied on a case by case basis.

   National Library of Australia
      The option is applied on a case by case basis.

   National Library of New Zealand
      The option is applied on a case by case basis.

   National Map Collection, PAC
      The option is applied on a case by case basis.

7B4   Variations in title. Make notes on titles borne by the item other than the title proper. If considered desirable, give a romanization of the title proper.   (3.7B4)

Panel title: Welcome to big Wyoming

Title in left margin: Ville de Aix-les-Bains, Savoie

Romanized title: Moskovskaiа oblast'

Panel title: Padre Island National Seashore, Texas

Cover title: Atlas regional, Chile VIII región

Vol. 1 issued in portfolio with title: Atlasul Republicii Socialiste
România

Running title: Economic atlas of Quebec

Title also in Chinese; romanized title: Hsiang-kang chieh tao yü ti ch'ü

Title on portfolio flap: Atlas régional du Nord-Pas-de-Calais

Title on spine of slip cases: France par départements

Spine title: Rocque's map of Shropshire

Spine title: Falk-plan Benelux

On spine: Atlas Jeune Afrique

Binder's title: Atlas minor 66 mapparum

Spine title (rebound v.): Secret maps of the Americas and the Indies from the Portuguese Archives

7B5 Parallel titles and other title information. Give parallel titles and other title information not recorded in the title and statement of responsibility area if they are considered to be important. (3.7B5)

Added title in Spanish

Subtitle on wrapper: Showing population changes 1951–60

■APPLICATION

Dedication. Many early cartographic items were dedicated by their authors to some influential person, explorer or patron.

A dedication not already transcribed in the title and statement of responsibility area may be transcribed in a note, preceded by the word *Dedication.*

Dedication: To the most serene and most sacred Majesty of Charles II. By the grace of God King of Great Brittain, France, and Ireland. This map of North America, is humbly dedicated

In a first-level description, parts of a lengthy dedication which do not contain essential information may be omitted. Indicate omissions from the dedication by the mark of omission (. . .).

In second- or third-level descriptions transcribe the dedication exactly.

7B6 Statements of responsibility. Make notes on variant names of persons or bodies named in statements of responsibility if these are considered to be important for identification. Give statements of responsibility not recorded in the title and statement of responsibility area. Make notes on persons or bodies connected with a work, or significant persons or bodies connected with previous editions, not already named in the description. (3.7B6)

Engraved by T.J. Newman

"Ch. Smith sculp."—Cover

"Plotted . . . by G. Petrie and D.P. Nicol, University of Glasgow, 1965. Field reconnaissance, 1962, and geomorphological interpretation by R.J. Price as part of project no. 1469 of the Institute of Polar Studies, the Ohio State University''

Attributed to Blaeu in: Atlantes Neerlandici / C. Koeman, vol. . . . , p. . . .

At head of title: Exxon Travel Club

''Form of the seismicity maps presented has been developed and designed by the Experimental Cartography Unit of the Natural Environment Research Council under Mr. D.P. Bickmore''

''Transcribed into Standard English Braille by Reggie Szava''—Cover

''Copyright, George Philip & Son, ltd.''—On each map
(*Of atlas*)

Maps drawn, engraved, and revised by Edward Weller and John Bartholomew

''Base map modified from Alabama Highway Department maps and field notes''

''Prepared by the Army Map Service (AM), Corps of Engineers, U.S. Army, Washington, D.C., from Lunar topographic map 1:5,000,000 Army Map Service, provisional edition, 1961''

Consultants: Joe B. Frantz and Henry F. Graff

Bathymetry compiled by the National Ocean Survey

7B7a Edition and history. Make notes relating to the edition being described or to the history of the cartographic item. (See 7A4 for form of notes.)   (3.7B7)

First ed. published 1954

Sheets of various eds.

A later state of the map first published in 1715 and in 1745. This state has the additions of ''King's roads'' and an advertisement for Overton's large map of the British Isles, dated 1746

Facsim. of: ''The 52 countries [sic] of England and Wales described in a pack of cards. Sold by Robert Morton . . . [et al.] in 1676''

The map plates, printed in Leipzig in English, derive from R. Andree's ''Allgemeiner Handatlas,'' 1887. The plates were later published in the first ed. of the ''Times atlas,'' 1895

Copied from:

Based on: ADAC Autoreisebuch Europa, and Guide de la route Europe

Red overprinting on the author's ''Greater Germany, administrative divisions 1 July 1944 (no. 3817-R&A, OSS)''

A later state of the map first published in 1772

From: Atlas élémentaire de géographie physique et politique / E.
Mentelle et P.G. Chanlaire. ₍1798₎

First ed. published as: Philips modern school atlas of comparative
geography. 1903

All previous eds. published by:

Differs from earlier edition by different typeface

Reprint. Originally published: Boston : Houghton Mifflin, 1910

Reprint of works originally published 1871–1929

''Reprinted in 1974 from the original copper plates in the possession of
the Hydrographic Office''

Published simultaneously as: McGraw-Hill international atlas. New
York : McGraw-Hill Book Co., 1963

■APPLICATION Give limited edition statements whenever possible, preferably in
quoted form. When the statement of limitation includes the unique number of the copy
being catalogued, give only the statement of limitation here. Give the copy number as a
copy-specific note (see 7B20).

Edition note: ''Special edition of 200 copies on handmade paper''
Copy-specific note: LC has no. 20, signed by author

As an alternative, give the entire statement of limitation and the copy number as a
copy-specific note.

7B7b Donor, source, etc., and previous owner(s). Make notes on the donor or
source of a cartographic item or a collection of cartographic items, and on previous
owners if they can be easily ascertained. Add the year or years of accession to the name
of the donor or source, and add the years of ownership to the name of a previous owner.

(4.7B7 mod.)

Gift of Worthington C. Ford, 1907

Purchase, 1951–1968

Purchased from the Del Monte collection, 1901

Gift of Mr. Wright, 1938–1954

Previously owned by L. McGarry, 1951–1963

Provenance: Bernard M. Bloomfield, 1904–   ₍Papers₎

7B7c Published versions of manuscript cartographic items. If the work contained
in a manuscript or the content of a manuscript collection has been, or is being, published, give the publication details.   (4.7B9 mod.)

Published as: The life of George Romney. London : T. Payne, 1809

Published in: Poetry : a magazine of verse. Vol. 59, 1942. p. 295– 308

Entire collection, with Jefferson papers from the Library of Congress and elsewhere, is being published in: Papers of Thomas Jefferson / edited by J.P. Boyd. 1950–

7B7d When it is known, give the location of, and other information about, the original of a reproduced cartographic item. (8.7B8 mod.)

Original in Prado Museum, Madrid

Original measures: 93 × 98 cm

Reproduced from a manuscript in the National Archives

Reproduced from copies in the Amsterdam University Library and in the Bibliotheca Thysiana, Leyden

An atlas factice of maps and text reprinted from various editions of: Grotten Atlas / Joan Blaeu. Amsterdam : J. Blaeu, 1630– 1662

7B8 Mathematical and other cartographic data. For celestial charts, give the magnitude.

Limiting magnitude 3.5

For remote-sensing imagery, give mathematical data not already included in the mathematical data area.

5.944, alt. 12,000 ft.

"Imagery recorded in discrete spectral bands with multispectral scanner (MSS) on NASA LANDSAT-1 (formerly ERTS-1). Orbital altitude 920 km. (570 mi.)"

Give other mathematical and cartographic data additional to or elaborating on that given in the mathematical data area. (3.7B8)

Scale of original: ca. 1:1 300

Oriented with north to right

Prime meridians: Ferro and Paris

Scale departure graph: "Statute miles Mercator projection"

Scale is correct for distances measured to and from Toronto only

"The 1:75,000 scale preliminary maps presented in this atlas were produced during the step reduction of 1:24,000 scale field maps"—Pref.

"All the relief-models are exactly the same scale as the corresponding political maps"
(From chief source)

■APPLICATION Grid and ellipsoid, etc., information on a cartographic item is recorded here.

Record a variation of orientation (north situated other than at the top of the sheet) if the variation is significant (e.g., 45 degrees or more from the top of the item).

> "Grid based on North Carolina rectangular coordinate system"

> "20,000-metre universal transverse Mercator grid, zones 16 and 17"

> International ellipsoid

> Military grid

> UTM grid

7B9 Publication, distribution, etc. Make notes on publication distribution, etc., details that are not included in the publication, distribution, etc., area and are considered to be important.   (3.7B9)

> Maps dated between 1780 and 1813

> The imprint of Gerard Valck has been substituted for the erased imprint of Joan. Blaeu, who probably first published the map ca. 1672

> Imprint of W. & S. Jones pasted onto the terrestrial and celestial globe gores

> Each map dated Feb., 1977

> Maps in pocket dated 1941–1943
> *(Imprint date 1933)*

> Date on cover label 1977

> Some sheets published by the U.S. Army Map Service

> Not distributed outside the U.S.

> "Published by arrangement with Evans brothers limited, London"
> *(Imprint:* Sydney : W. Standish & sons, [1941])

■APPLICATION   Also include significant differences in the date of publication or printing, significant statements of copyright, registration and/or deposit, etc. Include information on date of situation in 7B1.

If a manuscript cartographic item has an indication of the place in which it was done and this has not been transcribed in the description, record this and the source of the information in a note.   (4.7B8 mod.)

> At end: Long Beach Island

> On t.p.: London-Zagreb-Trieste

7B10a Physical description. Indicate any physical details that are considered to be important and have not been included in the physical description area.

                                                                    (3.7B10 mod.)

> Irregularly shaped

> Blue line print

Watermark: C. & I. Honig

In wooden case bearing, on its inner faces, representations of the celestial hemispheres

Bound in vellum

Maps dissected and pasted onto the sides of 42 wooden blocks to form an educational game

Free ball globe in transparent plastic cradle with graduated horizon circle and "geometer"

Pencil drawing

Pen and ink

Pen and ink and watercolour on tracing paper

Negative

Index map and distance chart on lining papers

Issued unbound in box

Each map preceded by transparent plastic overlay with boundaries in col. ink or crayon

Designed for use with radial index tape measure

Has grommets in upper corners for hanging

"This map is red-light readable"

7B10b Early printed cartographic items, signatures and foliation. Make a note giving details of the signatures and foliation of a volume.   (2.18D mod.)

Signatures: $a-v^8$, x^6

■APPLICATION   Give these details generally according to Gaskell's formula,[14] insofar as typographical facilities permit. Preface this note with the word *Signatures*.

Signatures: A^4 B$-$C^4 D^4 E$-$G^4 H^2

For incunabula, it is generally desirable to give either the signatures or a reference to a standard bibliographic source such as the *Gesamtkatalog der Wiegendrucke,* the *Catalogue of Books Printed in the XVth Century Now in the British Museum,* the *Catalogue general des incunables . . . / M.L.C. Pellechet, etc.,* as set out in 7B15.

If the gatherings are signed with one of the special characters used as abbreviation marks not within the capability of the typographical facilities available, substitute the spelled out form and enclose it in square brackets.

---

14. The following work is recommended as a standard for formulating these details: Gaskell, Philip. *A New Introduction to Bibliography.* — New York : Oxford University Press, 1972; or, Oxford : Clarendon Press, 1974. "Reprinted with corrections."

ᵣrumᵢ
ᵣetᵢ
ᵣconᵢ

If the gatherings are signed with other unavailable characters, substitute a descriptive term or an abbreviation for that term if a standard one exists.

ᵣdaggerᵢ
ᵣfleuronᵢ
ᵣpar.ᵢ
   (Gathering is signed with a paragraph mark: ¶)
ᵣsec.ᵢ
   (Gathering is signed with a section mark: §)

Whenever unsigned gatherings would be designated with the Greek letters pi and chi (cf. Gaskell, p. 330), substitute the roman alphabet form.

pi
chi

7B10c Reduction ratio for microforms. Give the reduction ratio if it is outside the $16\times - 30\times$ range. Use one of the following terms:

Low reduction	*For less than 16x*
High reduction	*For 31x−60x*
Very high reduction	*For 61x−90x*
Ultra high reduction	*For over 90x; for ultra high reduction give also the specific ratio, e.g.,* Ultra high reduction, 150×
Reduction ratio varies	

7B10d Reader. Give the name of the reader on which a cassette or cartridge microfilm is to be used if it affects the use of the item.   (11.7B10)

For Information Design reader

7B11 Accompanying material. Make notes on the location of accompanying material if appropriate. Give details of accompanying material not mentioned in the physical description area or given in a separate entry or separate description according to the rules for multilevel description (see 13F). (See also 5E1.)   (3.7B11)

Accompanied by filmstrip entitled: Mexico and Central America

Accompanied by the same maps in sheet form first published in: Géographie générale / M.J.C. Barbié Du Bocage. 1842

Each sheet accompanied by a sheet of geological sections

Accompanied by booklet: Djurberg, Daniel. Forklaring til Karten over Polynesien. Stockholm : Holmberg & Wennberg, 1780. 27 p. ; 20 cm. — Housed in Rare Books Collection

Accompanied by: Wissenschaftlicher Kommentarband

To accompany: Géographie physique / Nicolas Desmarest. (Encyclopédie méthodique; t. 100–104)

To accompany: New universal gazetteer / Cruttwell. 2nd ed. London, 1808

To accompany: Essai . . . 1812 which constitutes [pt. 3] of: Voyage de Humboldt . . . 1805–1834

Atlas suppl. to: Gesamtstudie über die Möglichkeiten der Fernwärmeversorgung aus Heizkraftwerken in der Bundesrepublik Deutschland, 1977

Companion atlas published separately: Atlas klimatyczny Polski

Suppl. to: Annals of the Association of American Geographers, v. 64, no. 4, Dec. 1975

"A supplement to the Catalogue of the Active Volcanoes of the World including Solfatara Fields"

"Devised as a companion to the Imperial Gazetteer"

"The atlas should be used in conjunction with Sand, gravel, and quarry aggregate resources, of the Colorado Front Range counties ; Colorado Geological Survey Special Publication 5-A"—Verso t.p.

"Complemented by the preceding bulletin, 'Seismicity and seismic hazard in Britain' by R.C. Lilwall (Seismological Bulletin no. 4)"

Errata slips and part of illustrative matter in pocket

Two folded col. maps in jacket pocket (21 × 9 cm)

Seven folded overlays in pocket

7B12 **Series.** Make notes on series data that cannot be given in the series area.

(3.7B12)

Original issued in series:

Some sheets have series designation:

"First in a series which will eventually cover the whole country"

"Contribution no. 1283-B, Division of Biology, Kansas Agricultural Experimental Station, Kansas State University"—Verso t.p.

"Contribution no. 658, Department of Botany, North Dakota Agricultural Experiment Station, North Dakota State University"—Verso t.p.

"A Philip/Salamander book"
   (Atlas)

Penguin reference books
   (Atlas)

Supersedes: General technical report RM-41
(Item being described is RM-78)

Sheets 8, 10, 12 marked: D.O.S. 30

7B13   Dissertations. If a cartographic item being described is a dissertation or thesis presented in partial fulfillment of the requirements for an academic degree, give the designation of the thesis (using the English word *thesis*) followed by a brief statement of the degree for which the author was a candidate (e.g., M.A. or Ph.D., or, for theses to which such abbreviations do not apply, *doctoral* or *master's*), the name of the institution or faculty to which the thesis was presented, and the year in which the degree was granted.   (2.7B13 mod.)

Thesis (Ph.D.)—University of Toronto, 1974

Thesis (M.A.)—University College, London, 1969

Thesis (doctoral)—Freie Universität, Berlin, 1973

If the publication is a revision or abridgement of a thesis, state this.

Abstract of thesis (Ph.D.)—University of Illinois at
Urbana-Champaign, 1974

If the thesis is a text edited by the candidate, include the candidate's name in the note.

Karl Schmidt's thesis (doctoral)—München, 1965

If the publication lacks a formal thesis statement, give a bibliographic history note.
                                                                 (2.7B13)

Originally presented as the author's thesis (doctoral—Heidelberg) under
the title:

7B14   Audience. Make a brief note of the intended audience for, or intellectual level of, an item if this information is stated in the item.   (3.7B14)

Intended audience: Primary schools

Access and literary rights. Indicate as specifically as possible all restrictions on the access to cartographic items.
   If the literary rights have been reserved for a specific period or are dedicated to the public and a document stating this is available, make an appropriate note.
                                                                 (4.7B14 mod.)

Accessible after 1983

Open to researchers under library restrictions

Information on literary rights available

Security classification

To be read in North Library only

"For official use only"

"Approved by the Ministry for Higher Education as a university textbook"

Restricted to U.S. Government officials

Unclassified

7B15   Reference to published descriptions. For incunabula, and *optionally* for other cartographic items, give the place in standard lists where the description of the item being described is to be found. Make this note in standard and abbreviated form.

(1.7B15, 2.18C mod.)

References: HR 6471; GW9101; Goff D-403

References: BMC (XV cent.) II, p. 346 (IB.5874); Schramm, v. 4, p. 10, 50, and ill.

Indexed by: Landownership maps, no. 696

Reference: Phillips 637

Reference: Similar to Phillips 4195

References: Hain 13541; BMC (XV cent.) IV, p. 133 (IC. 19313); Sabin 66474; Phillips 355

☐ POLICIES

British Library
   The option is applied on a case by case basis.

Library of Congress
   The option is applied whenever the edition being catalogued is listed in one of the following catalogues: Bristol, Blanck, Evans, Pollard and Redgrave, or Wing (use the forms shown here). It is also applied whenever the edition (including a facsimile) being catalogued is found in any other list or bibliography and the citation provides useful information (e.g., to distinguish editions, or to substantiate the cataloguer's conclusions).

National Library of Australia
   The option is applied whenever the edition being catalogued is the edition listed and information is readily available in a standard list. The number is quoted.

National Library of New Zealand
   The option is applied.

National Map Collection, PAC
   The option is applied.

Make a note on the best or fullest published description of a manuscript map or maps, atlas, etc., and published indexes or calendars.   (4.7B15 mod.)

> Calendar: Spanish manuscripts concerning Peru, 1531–1651. Washington, D.C. : Library of Congress, 1932

> Described and reproduced in: Portugaliae monumenta cartographica / A. Cortesão. Lisboa, 1960–₍1962₎, v. 4, p. 111–118, plates 464–472

7B16   Other formats available   (1.7B16)

> Also available on 3 × 5 cm colour slide

> Also available in 16 mm format

> Issued also in slip case

> Issued also on glass plates
> *(Sky atlas)*

7B18a   Contents.
If a collection of maps is described as a unit (see 0K), make notes on the state of the collection at the time of description and indicate the composition of the complete collection if possible. Give variations between sheets in the collection. Complete this note when the collection is complete.   (3.7B18)

> Complete in 174 sheets. Set includes various editions of some sheets including some reissued by the U.S. Army Map Service. Some sheets, prepared under the direction of the Chief of Engineers, U.S. Army, have series designation "Provisional G.S.G.S. 4145"

7B18b
Make notes describing the contents of an item (either partially or fully), including: components; insets; maps, etc., printed on the verso, or recto and verso of a map sheet, etc.; illustrations, etc. Give insets, etc., on the recto before maps, etc., on the verso of a sheet. Give the scale of insets, etc., if desired. If the insets, etc., are numerous and minor, give a note in general terms.   (3.7B18 mod.)

> Includes bibliographies

> Includes gazetteer, glossary, and index

> Includes key to 140 place names

> With two additional unnumbered parts: The stars in six maps. 1830 — The terrestrial globe in six maps. 1831

> Includes an index and illustrations of the Wangapeka Track

> Insets: Connaught Place — Chanakyapuri — Delhi & New Delhi City. Scale ₍ca. 1:23 000₎

> Insets: Political and economic alliances — Air distances from London — Membership of international organisations

> On verso: New map of South Hadley, Mass. Scale ₍ca. 1:15 000₎

Insets: Harrow ; Wembley ; Ruislip. On verso: Map of N.W. London

Includes 7 insets

Components: Ancient Orient before the rise of the Greeks. Scale
1:4 752 000 — Palestine about 860 B.C. Scale 1:506 880
(On an item with a collective title)

Components: The world in 3000 B.C. — The world in 1500 B.C. —
The world in 500 B.C. — The world in A.D.
(On an item with a collective title)

Components: Colonial organization of the world 1937 — Achievement
of independence 1958–1966
(On an item with a collective title)

Contents: Total dissolved solids in water from major streams —
Predominant chemical constituents in water from major streams —
Suspended-sediment concentration in major streams

Contents: Douglas Ranger District, Coronado National Forest
(Chiricahua-Peloncillo Mts.) Arizona and New Mexico — Douglas
Ranger District, Coronado National Forest (Dragoon Mts.) Arizona

Contents: Philips' atlas of ancient and classical history / edited by
Ramsay Muir and George Philip. 1938 — Muir's historical atlas,
mediaeval and modern. 7th ed. / edited by George Goodall. 1947
*(An atlas in one physical volume, consisting of two reprinted
bibliographic vols. Title proper:* Muir's atlas of ancient, mediaeval
and modern history)

Partial contents: Maarleveld, G.C. Geomorfologische waardering —
Maarleveld, G.C. Landschappelijke waardering — Kalkhoven, J.T.R.
Botanie — Bergh, L.M.J. van den. Ornithologie

Contents: [1] Kartenband — [2] Erläuterungsband

Contents: T. 1. Von km 865,5 (Deutsch-niederländ. Grenze) bis km
665,6 (Köln-Wesseling-Ndr.-Kassel) —

■ APPLICATION

1. Information from a cover or other container may be noted separately.

   Index on cover

2. Give all information concerning the recto before that of the verso. Put each of the
   different types of contents notes on separate lines or separate them from each
   other by a space, dash, space. Some of the more common types of contents notes
   are described below.

3. *Description of recto and verso.* Both *On recto* and *On verso* notes may be used
   to describe the elements of a sheet containing two or more maps printed on both
   sides, where an overall title for the sheet (found or supplied) has been used in the
   title area.

147

4. *Description of components.* Details of components are recorded in a description of components note, or a contents note.

    The components may be described collectively or individually in a sequence from the left of the map sheet to the right and from top to bottom, or in a sequential order (if so indicated on the item). Describe them as fully as desired.

    The description of components is always preceded by the word *Components* or *Contents*.

    Insets which are clearly insets of components may be described immediately following the component to which the inset belongs in as brief a manner as possible.

5. *Description of insets and ancillary maps.* Insets or ancillary maps may be described collectively or individually in a sequence from the left of the map sheet to the right and from top to bottom, or in a sequential order (if so indicated on the item). Describe them as fully as desired.

    Precede the description of insets or ancillary maps by the word *Inset(s)* or *Ancillary map(s)*.

    Insets which are clearly insets within insets may be described immediately following the inset to which it/they belong(s) in as brief a manner as possible.

6. *Description of illustrations, etc.* If illustrations, views, profiles, etc., are considered significant, describe them briefly. Also include information regarding text, indexes, etc., printed on the map sheet(s).

7. *Description of verso.* The *On verso* note contains the description of elements printed on the back of a map sheet. Describe any of these elements as fully as desired.

7B19   Numbers. Give important numbers borne by the item other than ISBNs or ISSNs (see 8B).   (3.7B19)

> Publisher's no.: LB 3721-9
>
> ''Plate no. 27''
>
> GPO stock no.: 024-005-00704-1
>
> ''Chart no. 694''
>
> United Nations sales publication no.: E.74.1.3
>
> Order no.: FRC M-63

■APPLICATION   Record important numbers, or alpha-numeric codes borne by the item other than those covered by series numbers, those covered by area 6, ISBNs and ISSNs. Give an explanatory phrase describing the type of number, if known.

7B20a   Copy being described and library's holdings. Make notes on any peculiarities or imperfections of the copy being described that are considered important. If the library does not hold a complete set of a multipart item, give details of the library's holdings. Make a temporary note if the library hopes to complete the set.   (3.7B20)

> Library's copy annotated in red ink to show land owners
>
> Library's copy imperfect: upper left corner missing
>
> Library's set lacks sheets 9–13 and sheet 27

"This is no. 16 of an edition limited to 100 copies"

LC has registered set no. 617

LC has no. 140

LC copy perforated with instruction for assembly as dymaxion globe; unassembled

7B20b   For early printed cartographic items make notes on special features of the copy in hand. These include rubrication, illumination, manuscript additions, binding, and imperfections.   (2.18F mod.)

Leaves I5– 6 incorrectly bound between h3 and h4

Imperfect: wanting leaves 12 and 13 (b6 and c1); without the blank last leaf (S8)

On vellum. Illustrations and part of borders hand col. With illuminated initials. Rubricated in red and blue

Contemporary doeskin over boards; clasp. Stamp: Château de la Roche Guyon, Bibliothèque

Blind stamped pigskin with initials C.S.A.C. 1644

Inscription on inside of front cover: Theodorinis ab Engelsberg

Signed: Alex. Pope

7B20c   Ancient, medieval, and Renaissance manuscripts. In addition to the notes specified above, give the following notes for ancient, medieval, and Renaissance manuscripts and collections of such manuscripts.

Illustrative matter. Give ornamentation, rubrication, illumination, etc., and important details of other illustrative matter.

Rubricated

Headings in red, with sepia drawings

Col. drawing of Jacob's dream on leaf $_{[}23_{]}$[a]

Collation. Give the number of gatherings with mention of blank, damaged, or missing leaves, and any earlier foliation.

Signatures (with catchwords at the end of each): $_{[}4_{]}$ leaves (on vellum), $_{[}a_{]}^{10}$, b^{10+2} (1st and last leaves on vellum), $c-f^{10}$, g^{10+2}, $h-p^{10}$, q^{10+2} (2nd and 11th leaves on vellum), $r-t^{10}$, v^8 (last 2 leaves blank)

Other physical details. Give details of owner's annotations, the binding, and any other important physical detail.   (4.7B22)

Annotated by previous owner, signed M.B.

Bound in calf, gold stamped, with Bellini arms on spine

Opening words. If the manuscript atlas is given a supplied title, quote as many of the opening words of the main part of the text as will enable the item to be identified.

(4.7B22 mod.)

> Tractatus begins (on leaf [17]ᵃ): Est via que videtʳ homī rcta nouissima . . .

7B21   "With" notes. If the description is of a separately titled part of a cartographic item lacking a collective title, make a note listing the other separately titled parts of the item in the order in which they appear there.   (3.7B21)

> With a separate map on same sheet: Queen Maud Range

> With (on verso): Motor road map of south-east England

> Mounted on a wooden stand to form a pair with: Bale's new celestial globe, 1845

> With: Atlas de France. Paris : Desnos, 1775

> With: Der Atlas des Andrea Bianco vom Jahre 1436

> With: Atlas elemental antiquo / por Don Tomás Lopez. Madrid, 1801

> Issued in portfolio with 19 other facsim. maps of New Jersey
> *(Other 19 maps catalogued separately, with same note)*

> Issued with: Glove compartment street atlas of Los Angeles / created and published by Goushā/Chek-Chart. San Jose, Calif. : Goushā/Chek-Chart, c1978
> *(One of two bibliographic works issued together in one physical volume without collective title; bound back-to-back/end-over-end, each with its own t.p., title, pagination, and catalogue record.)*

■APPLICATION   This note is only necessary when 1G6 is used.

8   STANDARD NUMBER AND TERMS OF AVAILABILITY AREA

Contents:

8A   Preliminary rule
8B   Standard number
8C   Key-title
8D   Terms of availability
8E   Qualification   (3.8)

8A   Preliminary rule   (3.8A)

8A1   Punctuation
For instructions on the use of spaces before and after prescribed punctuation, see 0C.
Precede this area by a full stop, space, dash, space *or* start a new paragraph.
Precede each repetition of this area by a full stop, space, dash, space.
Precede a key-title by an equals sign.

PICTORIAL

HACHURES

HACHURES

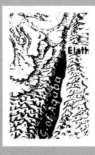
LANDFORM DRAWING

ROCK DRAWING

FORM LINES

CONTOURS

SPOT HEIGHTS

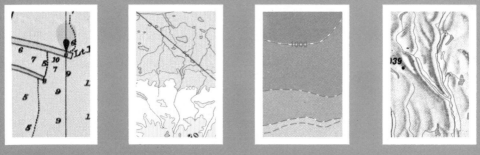
SOUNDINGS

HYPSOMETRIC TINTS

BATHYMETRIC TINTS

HILL SHADING

Figure 35   7B1   Examples of techniques for showing relief

Precede terms of availability by a colon.
Enclose a qualification to the standard number or terms of availability in parentheses.
(3.8A1)

8A2   Sources of information. Take information included in this area from any source.
Do not enclose any information in brackets.   (1.8A2)

8B   Standard number   (3.8B)

8B1   Give the International Standard Book Number (ISBN), or International Standard
Serial Number (ISSN), or any other internationally agreed standard number for the item
being described. Give such numbers with the agreed abbreviation and with the standard
spacing or hyphenation.   (1.8B1)

> ISBN 0-85152-392-7

> ISSN 0085-4859

8B2   If an item bears two or more such numbers, record the one which applies to the
whole item, or applies to the item being described.

Optionally, record more than one number and add a qualification as prescribed in 8E.
Give a number for a complete set before the number(s) for the part(s). Give numbers for
parts in the order of the parts.   (1.8B2)

> ISBN 0-379-00550-6 (set). — ISBN 0-379-00551-4 (v. 1)

☐POLICIES
> British Library
> The option is applied in all machine-readable records and selectively in
> printed records.

> Library of Congress
> The option is applied.

> National Library of Australia
> The option is applied.

> National Library of New Zealand
> The option is applied.

> National Map Collection, PAC
> The option is applied when the information is readily available to
> cataloguers.

8B3   If it is desired to include any number of an item other than an International
Standard Number, include such a number in a note (see 7B19).   (1.8B3)

8B4   If a number is known to be incorrectly printed in the item, give the correct
number if it can be easily ascertained and add *(corrected)* to it.   (1.8B4)

> ISBN 0-340-16427-1 (corrected)

8C Key-title (3.8C)

8C1 Add the key-title of a serial, if it is found on the item or is otherwise readily available, after the International Standard Serial Number (ISSN). Give the key-title even if it is identical with the title proper. If no ISSN is given, do not record the key-title.

(1.8C1)

8D *Optional addition.* **Terms of availability** (3.8D)

☐POLICIES

British Library
The option is applied only for records included in the UK MARC data base.

Library of Congress
The option is applied with few exceptions (e.g., price is omitted from the description of noncurrent items).

National Library of Australia
The option is applied.

National Library of New Zealand
The option is applied.

National Map Collection, PAC
The option is applied.

8D1 Give the terms on which the item is available. These terms consist of the price (given in numerals with standard abbreviations) if the item is for sale, or a brief statement of terms if the item is not for sale. (1.8D1)

: £4.40 (complete collection). — £0.55 (individual sheets)

: £2.50

: Free to students of the college

: For hire

■APPLICATION This information is usually recorded for current items only.

8E Qualification (3.8E)

8E1 Add after the standard number or terms of availability, as appropriate, a brief qualification when an item bears two or more standard numbers

ISBN 0-435-91660-2 (cased). — ISBN 0-435-91661-0 (pbk.)

ISBN 0-387-08266-2 (U.S.). — ISBN 3-540-08266-2 (Germany)

ISBN 0-684-14258-9 (bound) : $12.50. — ISBN 0-684-14257-0 (pbk.)
: $6.95

ISBN 3-411-01742-2 (Lw. in Schuber). — ISBN 3-411-01745-7 (Ldr. in Schuber)

and *optionally* when the terms of availability (see 8D) need qualification.

(1.8E1)

: £1.00 (£0.50 to members)

: $12.00 ($6.00 to students)

☐POLICIES

British Library
The option is applied only for records included in the UK MARC data base.

Library of Congress
The option is applied.

National Library of Australia
The option is applied.

National Library of New Zealand
The option is applied.

National Map Collection, PAC
The option is applied.

8E2   If there is no standard number, give the terms of availability before any qualification. (1.8E2)

$1.00 (pbk.)

9   **SUPPLEMENTARY ITEMS**   (3.9)

Contents:

9A   Describe supplementary items which are to be catalogued separately (see *AACR2* 21.28) as separate items. For instructions on the recording of the title proper of supplementary items, the titles proper of which consist of two or more parts, see 1B9.

(1.9A)

Atlas of sand, gravel, and quarry aggregate resources, Colorado Front Range counties / by S.D. Schwochow, R.R. Shroba, and P.C. Wicklin. — Scale 1:75 000. — Denver : Colorado Geological Survey, Dept. of Natural Resources, 1974
1 atlas (ca. 200 leaves) : chiefly maps ; 28 cm. — (Special publication / Colorado Geological Survey ; 5-B)
''The atlas should be used in conjunction with Sand, gravel, and quarry

aggregate resources, of the Colorado Front Range counties, Colorado
Geological Survey Special Publication 5-A''—Verso t.p.
> *(Each item, 5-A, and 5-B, has its own catalogue record, with access*
> *points under authors, title, and series; and with supplementary item*
> *note like the one above.)*

9B   Choose one of the following methods of describing supplementary items described
dependently: (1.9B)

1)   record the supplementary item as accompanying material (see 5E)

> 1 atlas (16 leaves) : chiefly maps ; 30 cm + legends

> maps : col. ; 53 × 79 cm or smaller, folded in cover 63 × 49 cm + legend
> ([1] leaf, 12 p. ; 61 cm)

or   2)   record minor supplementary items in the note area (see 7B11)

> Accompanied by supplement (37 p.) issued in 1971

> Accompanied by: The hydrology of the IFYGL Forty Mile & Oakville Creek
> study areas / by R.C. Ostry. Toronto : Ontario Ministry of the Environment,
> Water Resources Branch, 1979.   44 p.   (Water resource report ; 5b)

or   3)   use the multilevel description (see 13F).

10   ITEMS MADE UP OF SEVERAL TYPES OF MATERIAL   (3.10)

10A   This rule applies to items that are made up of two or more components, two or
more of which belong to distinct material types.   (1.10A mod.)

10B   If an item has one predominant component, describe it in terms of that component
and give details of the subsidiary component(s) as accompanying material following the
physical description (see 5E) or in a note (see 7B11).   (1.10B)

> 47 slides : col. ; 5 × 5 cm + 1 sound cassette

> 1 atlas ([13] leaves of braille and tactile graphics) : 6 col. tactile maps ;
> 28 cm + 1 sound cassette
> *Note:* Sound cassette title: Instructions, OSU Main Campus, tactual
> maps

10C   If an item has no predominant component, follow the rules below in addition to
the rules in the appropriate chapters of *AACR2*.   (1.10C)

10C1   General material designation. If general material designations are used (see
1C):

For an item without a collective title, give the appropriate designation after each title.
For an item with a collective title, follow the instructions in 1C4.   (1.10C1)

> . . . [kit]

> . . . [multimedia]

10C2   Physical description. Apply whichever of the following three methods is appropriate to the item being described.

a)   Give the extent of each part or group of parts belonging to each distinct class of material as the first element of the physical description (do this if no further physical description of each item is desired), ending this element with *in container,* if there is one, and following it with the dimensions of the container.

> 400 lesson cards, 40 answer key booklets, 1 student record, 1 teacher's handbook, 1 placement test, in container, 18 × 25 × 19 cm. — Lesson cards, arranged in 100 steps with 4 lessons at each level

or b)   Give separate physical descriptions for each part or group of parts belonging to each distinct class of material (do this if a further physical description of each item is desired). Give each physical description on a separate line.

> Beyond the reading list ₁multimedia₁ : guidelines for research in the humanities / C.P. Ravilious ; University of Sussex Library. — Brighton : University of Sussex Library, Audio-Visual Materials Room ₁distributor₁, 1975
> 46 slides : col.
> 1 sound cassette (15 min.) : 3 3/4 ips, mono.
> Summary: The bibliographic control of the humanities, with special reference to literature. A typical research project is followed through. — Intended audience: Postgraduates and research students

or c)   For items with a large number of heterogeneous materials, give a general term as the extent (see also 5B). Give the number of such pieces unless it cannot be ascertained.   (1.10C2)

> various pieces

> 27 various pieces

10C3   Notes. Give notes on particular parts of the item all together following the series area or following the physical description(s) if no series area is present.

(1.10C3)

10D   Multilevel description. In describing a single part of a multilevel item, follow the instructions in 13F.   (1.10D)

11   FACSIMILES, PHOTOCOPIES, AND OTHER REPRODUCTIONS   (3.11)

■APPLICATION   If a cartographic item is a facsimile, photocopy, or any other reproduction, check the scale in case the item has either been enlarged or reduced. If the item has been enlarged or reduced, compute the new scale from the bar scale if present. If the factor of reduction or enlargement is known, compute the scale from a verbal or reference fraction; if it is not known compute the scale by the comparison method. The scale of the original may be included in a note along with other details of the original (see 7B8).

11A In describing a facsimile, photocopy, or other reproduction of atlases, maps, manuscripts, and graphic items, give all the data relating to the facsimile, etc., in all areas except the note area. Give data relating to the original in the note area (but give numeric and/or alphabetic, chronological, etc., designations of serials in the material ₍or type of publication₎ specific details area). If a facsimile, etc., is in a form of library material different from that of the original (e.g., a manuscript reproduced as a book), use the *AACR2* chapter on the form of the facsimile, etc., in determining the sources of information (e.g., for a manuscript reproduced as a book, use *AACR2* chapter 2). In addition to instructions given in the relevant chapters, follow the instructions in this rule.   (1.11A mod.)

11B If the facsimile, etc., has a title different from the original, give the title of the facsimile, etc., as the title proper. Record the original title as other title information if it appears on the chief source of information of the facsimile, etc. (see 1D3). Otherwise, give the title of the original in the note area (see 11F).   (1.11B)

11C If the facsimile, etc., has the edition statement, publication details, or series data of the original as well as those of the facsimile, etc., give those of the facsimile, etc., in the edition, publication, distribution, etc., and series areas. Give the details of the original in the note area (see 11F).   (1.11C)

11D Give the physical description of the facsimile, etc., in the physical description area. Give the physical description of the original in the note area (see 11F).
                                                                              (1.11D)

11E If the facsimile, etc., has a standard number, give it in the standard number and terms of availability area, together with the key-title and terms of availability of the facsimile, etc. Give the standard number and key-title of the original in the note area (see 11F).   (1.11E)

11F Give all the details of the original of a facsimile, etc., in a single note. Give the details of the original in the order of the areas of the description.   (1.11F)

> Province of Manitoba and part of the District of Keewatin and North
> West Territory shewing the townships & settlements drawn from the latest
> gov. maps, surveys & reports for "The Prairie Province" / A.G.E.
> Westmacott. — Scale ₍ca. 1:1 363 000₎ (W 104°—W 94°/N 53°—N 49°).
> — ₍Ottawa₎ : Association of Canadian Map Libraries ₍and₎ Dept. of
> Geography, Brandon University, ₍197–₎. — 1 map ; 33 × 51 cm
> Facsim. of map in: Illustrated historical atlas of the County of Carleton.
> Toronto : H. Belden, 1879

13 ANALYSIS

Contents:

13C Note area
13D Analytical added entries
13E ''In'' analytics
13F Multilevel description   (13)

13A Scope

Analysis is the process of preparing a bibliographic record that describes a part or parts of a larger item. The following rules offer various ways of achieving analysis. Some of these methods of analysis are related to provisions found in previous rules, but all the methods are collected here with general guidelines to assist in the selection of one of the means of analysis. Cataloguing agencies have their own policies affecting analysis; in particular, a policy predetermining the creation of separate bibliographic records may override any other consideration.   (13.1)

13B Analytics of monographic series, map series, and multipart cartographic materials[15]

If the item is part of a monographic series, a map series (see Glossary), or a multipart cartographic item and has a title not dependent on that of the comprehensive item, prepare an analytical entry in terms of a complete bibliographic description of the part. Give details of the comprehensive item in the series area (see 6).   (13.2 mod.)

> Motorists' map of northern Scotland / designed and produced by the Cartographic Unit (Publications Division) of the Automobile Association. — 1st ed. — Scale 1:316 800. 5 miles to 1 in. — Basingstoke, Hampshire : A.A., c1977. — 1 map : both sides, col. ; 80 × 104 cm, on sheet 42 × 116 cm. — (AA 5 miles to 1 inch easy-fold series : [Great Britain] ; sheet 6). — ISBN 0 09 211310 9

13C Note area

If a comprehensive entry for a larger work is made, this entry may contain a display of parts in the note area (normally in a contents note). This technique is the simplest means of analysis; the bibliographic description of the part is usually limited to a citation of title or name and title.   (13.3)

> Roadline primary route map Great Britain / [by] Geographia Ltd. — Scale 1:300 000 [i.e. 1:316 800]. 5 miles to 1 in. 4 km to 1 cm. — London : Geographia [ca. 1971]. — 6 maps : col. ; 74 × 101 cm. — Contents: 100 miles around London: sheet 1 — South west England and south Wales: sheet 2 — South Wales and the Midlands: sheet 3 — Northern England: sheet 4 — Central and southern Scotland: sheet 5 — Northern Scotland: sheet 6

■APPLICATION The bibliographic description of each part in the note area should contain as a minimum title, sheet number, series number (if appropriate), and scale (if this is more precise or differs from the entry in the main body of the description).

---

15. For the treatment of map series see Appendix E.

13D Analytical added entries

If a comprehensive entry for a larger work is made that shows the part either in the title and statement of responsibility area or in the note area, an added entry for the part may also be made. The heading for this added entry consists of the part's main entry heading plus uniform title (see *AACR2* 21.30M). This method is appropriate when direct access to the part is wanted without creating an additional bibliographic record for the part. (13.4)

13E "In" analytics

If more bibliographic description is needed for the part than can be obtained by displaying it in the note area, the "In" analytic entry may be considered. (13.5)

13E1a The descriptive part of an "In" analytic entry consists of a description of the part analyzed followed by a short citation of the whole item in which the part occurs.

The description of the part analyzed consists of:

i)	elements of the title and statement of responsibility area that apply to the part
ii)	elements of the edition area that apply to the part
iii)	numeric or other designation, if a serial (mathematical data area)
iv)	elements of the publication, distribution, etc., area, if the part is a monographic item and if the elements differ from those of the whole item
v)	extent and specific material designation of the part (when appropriate, in terms of its physical position within the whole item)
vi)	other physical details
vii)	dimensions
viii)	notes. (13.5A mod.)

13E1b The citation of the whole item (the "In" analytic note) begins with the word *In* (italicized, underlined, or otherwise emphasized) and consists of the main entry heading and uniform title (in square brackets) as appropriate (see *AACR2* Part II) of the whole item, together with its title proper, edition statement, numeric or other designation (of a serial), or publication details (of a monographic item). (13.5A)

■APPLICATION   According to 13E1a5 either the extent and the specific material designation of the part, or the location of the part within the whole item is recorded in the physical description. However, it is preferable to record the extent and specific material designation of the part in the physical description. Record the location of the part within the whole item in a note.

> Plan shewing the region explored by S.J. Dawson and his party
> between Fort William, Lake Superior and the Great Saskatchewan River,
> from 1st of August 1857 to 1st November 1858. — Scale ₍1:633 600₎ (W
> 105°—W 87°/N 54°—N 48°). — Toronto ₍C.W.₎ : J. Ellis, ₍1859?₎
> 1 map ; 112 × 203 cm, folded to 28 × 23 cm
> *In* Dawson, S.J. Report on the exploration of the country between Lake
> Superior and the Red River settlement. — Toronto ₍C.W.₎ : John Lovell,
> Printer, 1859

> Moral & statistical chart shewing the geographical distribution of man
> according to religious belief, with the principal Protestant mission stations

in the middle of the 19th century / by A. Keith Johnston. — Scale
₍1:65 000 000₎ (W 180°—E 180°/N 77°—S 56°). — Edinburgh ; London :
William Blackwood & Sons, 1st May 1854 (Edinburgh : W. & A.K.
Johnston)

    1 map : col. ; 36 × 60 cm, on sheet 56 × 67 cm, folded to 56 × 34 cm
    In upper margin: Phytology & zoology no. 11
    Insets: Mission stations of British N₍or₎th America — Mission stations
of South Africa — Distribution of man in Europe according to language
— Mission stations of India — Prevailing forms of religion in Europe —
3 diagrams
    In Johnston, Alexander Keith. The physical atlas of natural
phenomena.— A new and enl. ed. — Scales differ. — Edinburgh :
William Blackwood and Sons, 1856. Plate 34
       (Imprint on map and on atlas)

    The world / comp. and drawn in the Cartographic Division of the
National Geographic Magazine. — Scale 1:39 283 200. 620 miles to the
in. at the equator ; Van der Grinten proj. (W 180°—E 180°/N 84°—S
74°). — Washington, D.C. : National Geographic Society, 1970

    1 map : col. ; 68 × 102 cm, folded to 16 × 24 cm
    Insets: ₍Antarctica₎. Scale ₍ca. 1:50 000 000₎ — ₍North Polar region₎.
Scale ₍ca. 1:50 000 000₎ — The United Nations — Population density —
Vegetation and land use — International time zones
    On verso, poster: How man pollutes his world
    In Exploring the world of maps. — Washington, D.C. : National
Geographic Society, c1973
    5 filmstrips, 5 sound cassettes, 5 booklets, 8 maps
       (The maps in this kit have all been published previously by the
       National Geographic Society and each bears its own publication
       data. The entry for the component part comes completely from the
       part and the entry for the whole item comes from the container.)

13E2   Parts of "In" analytics

If an "In" analytic entry is made for a part of an item that is itself catalogued by
means of an "In" analytic entry, the "In" analytic note contains information about the
whole item and about the part containing the part being analyzed. Give information
about the smaller item first, and then information about the comprehensive item in the
form of a series statement.   (13.5B)

    Canadian-owned paper industries = Industrie des pâtes et papiers de
propriété canadienne / comp. by D. Michael Ray ; cartography by Policy
Research and Coordination Branch. — Scale ₍1:95 000 000₎ (W 134°—W
52°/N 57°—N 42°). — ₍S.1. : s.n., 1971₎ (Ottawa : Surveys & Mapping
Branch, 1970)

    1 map : col. ; 25 × 58 cm, folded to 28 × 20 cm
    Title in title block: Canada
    Inset: ₍Montreal region₎
    In Ray, D. Michael. Dimensions of Canadian regionalism. — Ottawa :

Policy Research and Coordination Branch, Dept. of Energy, Mines and
Resources, 1971. — (Geographical paper ; no. 49). Figure 26

13F   Multilevel description[16]

Multilevel description is normally used by national bibliographies and those cataloguing agencies that prepare entries needing complete identification of both part and
comprehensive whole in a single record that shows as its primary element the description of the whole. It may sometimes be used as an alternative to "In" analytic entries.

Divide the descriptive information into two or more levels. Record at the first level
only information relating to the multipart item as a whole. Record at the second level
information relating to a group of parts or to the individual part being described. If
information at the second level relates to a group of parts, record information relating to
the individual part at a third level. Make the levels distinct by layout or typography.

> AA 5 miles to 1 inch easy-fold series : ₍Great Britain₎. — Scale
> 1:316 800. 5 miles to 1 in. — Basingstoke, Hampshire : A.A., ₍1977₎– .
> — 6 maps on 1 sheet : both sides, col. ; sheet 42 × 116 cm
>   Sheet 6: Motorists' map of northern Scotland / designed and produced
> by the Cartographic Unit (Publications Division) of the Automobile
> Association. — 1st ed. — c1977. — ISBN 0 09 211310 9

> Roadline primary route map Great Britain / ₍by₎ Geographia Ltd. —
> Scale 1:300 000 ₍i.e. 1:316 800₎. 5 miles to 1 in. 4 km to 2 cm. —
> London : Geographia, ₍ca. 1971₎– . — 6 maps : col. ; 74 × 101 cm
>   Sheet 2: South west England and south Wales. — ₍ca. 1978₎. — Date
> code L.LA CU.LA believed to be 7.78 12.78. — ISBN 0 09 203470 5

Complete any element left open when all the parts are received.   (13.6)

■APPLICATION   When recording a title of a part in a multilevel description, any numbering is given first and the title is separated from it by colon, space. Access points may
be made for entries at the second (and subsequent) levels.

Title proper. Record the title proper, which is common to all sheets, at the first level.
Give the sheet title at the second level; do not indicate its omission at the first level by
the mark of omission.

General material designation. When used, give the general material designation at
the first level.

Statements of responsibility. When the statement of responsibility is common to all
sheets, give it at the first level. When the statement of responsibility is specific to a
particular sheet, record it at the second level for that sheet.

Edition statement. When a cartographic work is entirely republished, record the edition statement at the first level. When a sheet is republished, give the edition statement
at the second level.

Mathematical data. When the mathematical data are the same for all sheets, give

---

16. For the treatment of map series see Appendix E.

them at the first level. When the mathematical data vary from sheet to sheet, give them at the second level.

Publication, distribution, etc. Generally record the place of publication, distribution, etc., and the name of the publisher, distributor, etc., at the first level.

If the name of the publisher, distributor, etc., changes over the course of the publishing history, record the earliest form of the name that occurs at the first level. At the second level record the name as given on the item.

If the publisher changes over the course of publication and there are three or fewer, record all of them at the first level. If there are more than three publishers record at the first level [Various places : various publishers], or [*place*] : [various publishers]. Record the publisher(s) as given on the item at the second level.

For a work published cooperatively where each body is responsible for only part of the whole, record at the first level three or fewer names of places, publishers, etc. (see 4B8). If there are four or more, record [Various places : various publishers]. Record the publisher of the individual sheet at the second level.

Date of publication. When all sheets of a cartographic item have been published, give the first and last dates at the first level. Give the dates of each sheet at the second level. Record an open date at the first level for a cartographic item that is in the course of publication (see 4F8).

When the place of printing, name of the printer, and date of printing for the complete cartographic item (all sheets) are the same, record them at the first level. When the place of printing, the name of the printer, and the date of printing are all different, or either one is different, record them (it) at the second level.

Record the place of printing, the name of the printer, and the date of printing at the second level when the cartographic item is in the course of publication.

Physical description. Record the specific material designation at the first level. Give the statement of extent at the first level when all sheets of the cartographic item have been published and described. In all other cases, record the statement of extent at the second level.

Other physical details and dimensions may be given at both levels provided that the total physical description area results in an intelligible statement at each level.

Notes. Record notes pertaining to the entire cartographic work (all sheets) at the first level. Give notes pertaining to individual sheets at the second level for each sheet.

Standard number and terms of availability. Record all elements in this area at the first level if they pertain exclusively to the entire cartographic work.

APPENDICES

A   Guidelines for Choice of Access Points
B   Guidelines to Determine the Scale (RF) of a Map
C   Date of Situation
D   Series
E   Treatment of Map Series
F   Geographical Atlases
G   Examples
H   Abbreviations
J   Numerals

Appendix **A**

Guidelines for Choice of Access Points

A.0   INTRODUCTION

AACR2 rule 20.1 specifically states that the rules in Part II apply to all library material, irrespective of the medium in which they are published or of whether they are serial or nonserial in nature. However, application of the rules in *AACR2* chapter 21 to cartographic materials is complicated by differences[1] between them and other media.

■APPLICATION   In particular these differences may result in difficulties in determining whether a personal author or corporate body is chiefly responsible for the creation of the intellectual and artistic content of the cartographic work. If a personal author such as defined in A.1 Application cannot be distinguished, enter under corporate body (A.2) if appropriate. If neither heading is applicable enter under title (*AACR2* 21.1C).

---

1. Differences occur in terms such as compiler, series, etc.; differences in design, arrangement, and scope of bibliographic information; and, differences in publication methods.

A.1 Definition

A personal author is the person chiefly responsible for the creation of the intellectual or artistic content of a work. For example, writers of books and composers of music are the authors of the works they create; compilers of bibliographies are the authors of those bibliographies; cartographers are the authors of their maps; and artists and photographers are the authors of the works they create. In addition, in certain cases, performers are the authors of sound recordings, films, and videorecordings. (21.1A1)

■APPLICATION Read "cartographers may be the authors of their maps" instead of the statement "cartographers are the authors of their maps." It is better to state this particular class of personal authorship of cartographic materials as only a possibility since the intellectual responsibility of cartographers can vary from complete responsibility for the cartographic item as a whole to only partial responsibility that may in some cases be no more than photographic or mechanical duplication or tracing of an existing base map.

A.2 Add the following category to those listed under *AACR2* 21.1B2:

f) cartographic materials emanating from a corporate body other than a body which is merely responsible for the publication and distribution of the material.

■APPLICATION The emanating corporate body is responsible for the intellectual content, design, and creation of the material. However, the name of the body is not always printed prominently on an item and is not always accompanied by an explicit statement of responsibility. Where there is difficulty in determining the degree of responsibility for the cartographic item that such a corporate body may have had, consider what is known about the publication history of the body and if it is known to be a map-making organization which normally originates and issues maps, enter under the corporate body. If the principal responsibility for the production of the cartographic item is clearly attributed to a personal author, do not consider that this category applies.

Appendix **B**

Guidelines to Determine the Scale (RF) of a Map

Contents

B.1 Introduction
B.2 Methods of determining a representative fraction (RF)
B.3 Index to tables. Tables

B.1 INTRODUCTION

These guidelines and tables are provided as a handy reference. The cataloguer must be aware of the limitations in accuracy of the methods used to derive a representative fraction (RF). For many older maps the stated or computed scale may be notional as these maps were often constructed and drawn in such a way that the scale varies erratically across the face of the map according to the data available to the maker. For most modern maps it is the projection which is the most important element affecting the variation or distortion of scale or scales over the surface of the map. It must be thoroughly understood, however, that no flat map has a constant scale over its entire surface. The most accurate measurements can be obtained in the central area of the map (see B.2D).

On small-scale maps (those covering large areas) the properties of the projection affect the scale variation and distortion over the surface of the map. Thus the smaller the scale, the larger the distortion and scale variation.

On larger scale maps (those covering smaller areas) less distortion occurs and hence there is less scale variation from place to place over the surface of the map. This is due to the reduced amount of "squeezing" required to map a small area of the curved surface of the earth on flat paper. Thus on large-scale maps a RF can be calculated from distances measured in any direction from any point. The resulting scale from calculations from these data will be accurate enough for cartobibliographical purposes.

Scales on cartographic items are given in three basic forms, and may appear there in any combination of these forms as follows:

a) a graphic scale representation or bar scale (see figure B1)

b) a verbal scale in the form of a phrase or sentence stating map distance in relation to earth distance, or earth distance in relation to map distance

 2 cm to 1 km

 1 inch to 1 mile, or 1 mile to the inch

c) a representative fraction (RF).

 1:50 000

 1:63 360

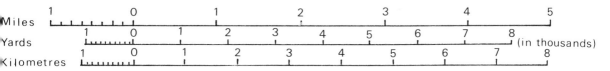

Figure B1   Graphic scales

B.2   METHODS OF DETERMINING A REPRESENTATIVE FRACTION (RF)

The representative fraction may be determined by use of a *natural scale indicator*[1] for those cartographic items with parallels of latitude or a graphic scale of linear distance in miles, feet, kilometres, etc.

Four methods of determining a representative fraction are explained below in their order of priority:

---

1. Currently available (1982) is a Scale Indicator from the Memorial University of Newfoundland Cartographic Laboratory, St. John's, Newfoundland, Canada A1B 3X9.

A. conversion from scale statements in the form of a phrase
B. conversion from graphic scales
C. determination of the RF of a map of unknown scale by using its graticule
D. determination of the RF by means of comparison with a map of known scale (comparison method).

B.2A Conversion from scale statements in the form of a phrase

Most such statements express in verbal form the ratio of units on the map to units on the ground, or vice versa.

> 1 inch to 2 miles or 2 miles to the inch
>
> 1 cm for 2 km
>
> 2½ inches to the mile
>
> ¼ inch to the mile
>
> 2 640 feet to the inch

When a map is photocopied or otherwise reproduced, the scale often differs from that of the original. Thus a scale statement expressed in a verbal form on the photocopied item will be inappropriate. The representative fraction should be determined by one of the other methods outlined here.

To convert the verbal scale statements to RF form, the units on one side of the ratio are converted to the same units on the other side, with the units on the map (numerator of the fraction) expressed as unity ($= 1$).

1. 1 inch to 2 miles or 2 miles to the inch

    1 inch (on map) represents 2 miles (earth)
    2 miles = 2 × 63 360 inches
    The ratio is: 1:2 × 63 360
                        or
                    1:126 720

2. 1 cm for 2 km

    1 cm (on map) represents 2 km (earth)
    2 km = 2×10^5 cm
    The ratio is: 1:2 × 10^5
                        or
                    1:200 000

3. 2½ inches to the mile

    2½ inches (on map) represents 1 mile (earth)
    1 mile = 63 360 inches
    The ratio is: 2½:63 360
                        or
                    $1:\dfrac{63\ 360}{2½}$

or
1:25 344

4.   ¼ inch to the mile

¼ inch (on map) represents 1 mile (earth)
1 mile = 63 360 inches
The ratio is: ¼:63 360

or

$$1:\frac{63\ 360}{¼}$$

or
1:253 440

5.   2 640 feet to the inch

1 inch (on map) represents 2 640 feet (earth)
2 640 feet = 31 680 inches
The ratio is: 1:31 680

B.2B   Conversion from graphic scales[2]

Graphic scales are most commonly given in the form of a bar scale. (See Figure B2.)

To determine the RF from a bar scale, the bar is measured with a graduated ruler or a natural scale indicator (see Figure B2). Although it is not absolutely necessary, it is advisable to use a ruler with the same system of measurement as that of the bar scale.[3] For example, if the bar scale is in metric units, use a metric ruler.

If the units along the bar are in an obsolete system of measurement which is unknown, or no conversion factor can be found for it,[4] another method of RF determination must be applied.

If a natural scale indicator is used, normally the scale can be read from it directly. Any calculations that may be necessary are included in the instructions accompanying the indicator.

The bar scale is measured in the units of the graduated ruler, then that number of units (measured) on the bar is converted to the same units measured by the graduated ruler. The ratio or RF is the ratio of the units measured by the graduated ruler to the

---

2. Calculations from measurements of a bar scale on the map should be restricted as much as possible to large-scale maps and plans (i.e., larger than 1:50 000), because the limitations in accuracy inherent in measuring the bar itself with a graduated ruler results in gross inaccuracies. The smaller the scale, the larger the inaccuracy (see example 2 of B.2B and footnote).

3. The practice of using a graduated ruler with units of the same system of measurement in which the bar scale is divided is not necessary for the calculation of a RF. However, scale calculations are simpler when one system of measurement is used. Such calculations become more complex when it is necessary to convert from one system of measurement to another system. *See* International Geographical Union. *Geographical Conversion Tables* / compiled and ed. by D.H.K. Amiran and A.P. Schick. — Zurich : I.G.U., 1961.

4. For conversion figures for some obsolete systems of measurement see: Doursther, Horace. *Dictionnaire universel des poids et mesures anciens et modernes, contenant des tables des monnaies de tous les pays.* — Amsterdam : Meridian Publishing Co., 1965. Or *see* Johnstone, William D. *For Good Measure : a Complete Compendium of International Weights and Measures.* — 1st ed. — New York : Holt, Rinehart and Winston, c1975. — Contains units of length and area for both ancient and modern countries.

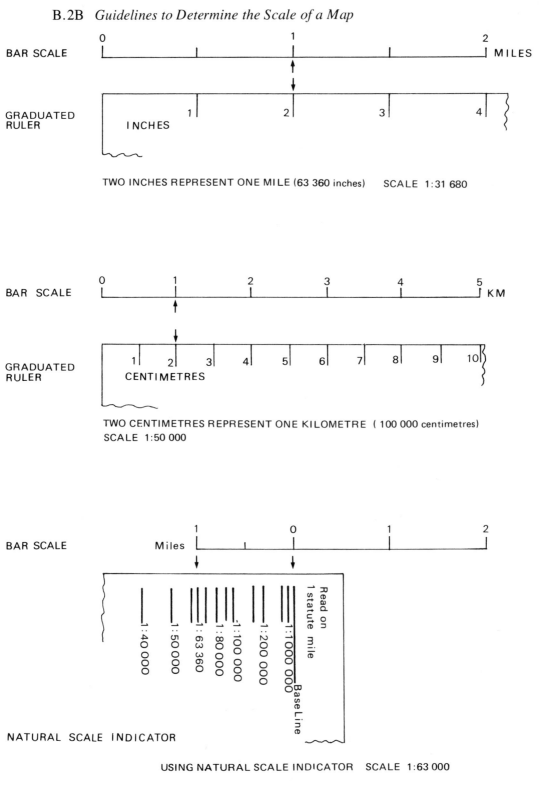

Figure B2   Graphic scales

units (measured) on the bar. This ratio represents units on the map to units on the surface (real) of the earth, where the units on the map (or the numerator of the fraction) is always expressed as unity (= 1). The ratio can be written as a fraction, e.g., $\frac{1}{63\ 360}$ or 1/63 360 but is most commonly found written 1:63 360 or 1:63,360.

1. 2 miles in units of the bar scale measure 4 inches with a ruler.

   4 inches (on map) represent 2 miles (earth)
   1 mile contains 63 360 inches
   2 miles = 2 × 63 360 inches
   The ratio is: 4:(2 × 63 360)

   or

   $$1:\frac{2 \times 63\ 360}{4}$$

   or
   1:31 680

2. 2 000 yards in units of the bar scale measure 1.14 inches with a ruler.

   1.14 inches (on map) represents 2 000 yards (earth)
   2 000 yards = 2 000 × 36 inches
   The ratio is: 1.14 :(2 000 × 36)

   or

   $$1:\frac{2\ 000 \times 36}{1.14}$$

   or
   1:63 157[5]

3. 50 chains in units of the bar scale measure 1½ inches with a ruler.

   1½ inches (on map) represents 50 chains (earth)
   1 chain contains 22 yards or 22 × 3 × 12 inches
   50 chains contain 50 × 22 × 3 × 12 inches
   The ratio is: 1.5:50 × 22 × 3 × 12

   or

   $$1:\frac{50 \times 22 \times 3 \times 12}{1.5}$$

   or
   1:26 400

4. 5 kilometres (km) in units of the bar scale measure 10 centimetres (cm) with a ruler.

---

5. The example illustrates the inherent limitations of accurate measurement when this method is used for small-scale maps. In this case 1.14 inches was measured with a ruler. The accuracy required should have been to 10[5] of an inch, namely 1.13636 inches as the map is actually at a scale of 1:63 360 or 1 inch to the mile, i.e.,

$$1:\frac{2\ 000 \times 36}{1.13636} = 1:63\ 360$$

10 cm (on map) represents 5 km (on earth)

1 km contains 100 000 cm or 10^5 cm

5 km $= 5 \times 10^5$ cm

The ratio is: $10:5 \times 10^5$

or

$1:5 \times 10^4$

or

1:50 000

5. 1 000 toises (French measure before 1793) in units of the bar scale measure 28.1 cm with a ruler.

28.1 cm (on map) represents 1 000 toises (earth)

1 toise $= 1.949$ metres

1 000 toises $= 1\,000 \times 1.949 \times 10^2$ cm

$= 1.949 \times 10^5$

The ratio is: $28.1:1.949 \times 10^5$

or

$$1:\frac{1.949 \times 10^5}{28.1}$$

or

1:6 936

B.2C Determination of RF of a map of unknown scale by using its graticule

The following formula can be used to calculate an approximate RF:

$$\frac{n}{11\,000\,000}$$

where,

n   is the number of centimetres measured on the map for one degree of latitude

11 000 000   represents the approximate number of centimetres on the ground for one degree of latitude.

$$\frac{44}{11\,000\,000} = \frac{1}{250\,000} = 1{:}250\,000$$

$$\frac{26.5}{11\,000\,000} = \frac{1}{415\,000} = 1{:}415\,000$$

B.2D Determination of the RF by means of comparison with a map of known scale (the comparison method)

This method provides a useful check for RF values obtained by other means. It can be used to obtain a RF for a map which has no indication of scale and for which no scale can be determined from a secondary source. This method can always be applied if all other means of scale determination fail.

It must be realized that this method is one of the least accurate ways of calculating RF and background knowledge of cartography is necessary for its correct application. The method involves an actual scale calculation from measurements taken on the map itself. The accuracy of the calculation depends to a large extent on:

1. the scale of the map of known scale on which a distance is measured
2. the projection of that map and the map of unknown scale
3. the location and direction from which distances are measured.

Such measurements will be more accurate if taken from the central area of the map. They should be taken east-west along the parallels, or north-south along the meridians. If the map projection has one or more standard parallels, measurements taken along those parallels will be true to scale. For example, early chart measurements along the north-south trending coastlines are more accurate than east-west measurements.

The method is based on obtaining the true (ground) distance between two identifiable points on the map. This true (ground) distance is obtained by measuring from a map of known scale and accuracy. Once the true ground distance is known between the two identifiable points on the map of unknown scale, use the following formula:

$$n = \frac{m \times a}{b}$$

where,

a. n   is the denominator of the scale fraction of map with unknown scale
m   is the denominator of the scale fraction of map with known scale
a   is the distance measured on the map with the known scale between two identifiable points
b   is the distance measured on the map with the unknown scale between the same identified points.

Note: a and b have to be in the same units.

1. The distance between two identifiable points on map of unknown scale measures 5 cm. The distance between the same points on a map at a scale of 1:50 000 measures 25 cm.

   Applying the formula: $n = \frac{50\ 000 \times 25}{5}$

   $n = 250\ 000$

   The unknown scale is 1:250 000.

2. The distance between two identifiable points on map of unknown scale measures 29.6 inches. The distance between the same points on a map at a scale of 1:50 000 measures 37.5 inches.

   Applying the formula: $n = \frac{50\ 000 \times 37.5}{29.6}$

   $n = 63\ 345$

   The unknown scale is 1:63 345.

   Some further checking with other methods of determination of the RF showed that the map was at a scale of 1:63 360 or 1 inch to the mile.[6]

---

6. Again it must be stressed that no flat map has a constant scale over its entire surface. Scales are usually only true at a point, or along a certain line (meridian or parallel) on the map.

B.3 *Guidelines to Determine the Scale of a Map*

To check whether any gross inaccuracy is made, it is wise to determine the RF by several methods or to identify and measure two distances preferably in different directions and averaging the RFs thus obtained.

B.3 INDEX TO TABLES

TABLE 1.   Conversion within the British System

Unit	yards	feet	inches
Cable	240	720	8 640
Chain	22	66	792
Fathom	2	6	72
Furlong	220	660	7 920
League (3 miles)	5 280	15 840	190 080
Link		0.66	7.92
Mile, Nautical, British	2 026.75	6 080.27	72 963.24
Mile, Nautical = International Nautical Mile	2 025.37	6 076.1[a]	72 913.2
Mile, Statute	1 760	5 280	63 360
Pace			30 approx.
Perch ⎫			
Pole ⎬	5.5	16.5	198
Rod ⎭			
Acre	4 840²	43 560²	

a. Before 1 July 1954 one U.S. Nautical Mile was 6 080.21 ft.

TABLE 2.   Conversion within the Metric System

	mm	cm	m	km
1 mm =	1	0.1	0.001	0.000 001
1 cm =	10	1	0.01	0.000 01
1 m =	1 000	100	1	0.001
1 km =	1 000 000	100 000	1 000	1

TABLE 3.   Greenwich Longitude of Various Prime Meridians

Amersfoort, Netherlands	5°	23'		E
Amsterdam, Netherlands	4°	53'	01"	E
Antwerp, Belgium	4°	22'	50"	E
Athens, Greece	23°	42'	59"	E
Berlin, Germany	13°	23'	55"	E
Bern, Switzerland	7°	26'	22"	E
Bogotá, Colombia	74°	04'	53"	W
Bombay, India	72°	48'	55"	E
Brussels, Belgium	4°	22'	06"	E
Bucharest, Romania	26°	07'		E
Cádiz, Spain	6°	17'	42"	W
Canberra, Australia	149°	08'		E
Capetown, South Africa	18°	28'	41"	E
Caracas, Venezuela	66°	55'	50"	W
Celebes, Middle Meridian of, Indonesia	121°	48'		E
Copenhagen, Denmark	12°	34'	40"	E
Córdoba, Argentina	64°	12'	03"	W
Djakarta (Batavia), Indonesia	106°	48'	28"	E
Ferro (Hierro), Canary Islands	17°	39'	46"	W[1]
Genoa, Italy	8°	55'		E
Helsinki, Finland	24°	57'	17"	E
Istanbul, Turkey	28°	58'	50"	E
Julianehaab, Greenland	46°	02'	22"	W
Leningrad, U.S.S.R.	30°	18'	59"	E
Lisbon, Portugal	9°	11'	10"	W
London, U.K.	0°	05'	43"	W
Madras, India	80°	14'	50"	E
Madrid, Spain	3°	41'	15"	W
Mexico City, Mexico	99°	11'	40"	W
Moscow, U.S.S.R.	37°	34'	15"	E
Munich, Germany	11°	36'	32"	E
Naples, Italy	14°	15'	42"	E
Oslo (Christiania), Norway	10°	43'	23"	E
Padang, Sumatra, Indonesia	100°	22'	01"	E
Paris, France	2°	20'	14"	E
Peking, China	116°	28'	10"	E
Philadelphia, Pa., U.S.A.	75°	08'	55"	W
Pulkovo[2] (Leningrad), U.S.S.R.	30°	19'	39"	E
Quito, Ecuador	70°	30'		W
Rio de Janeiro, Brazil	43°	01'	21"	W
Rome, Italy	12°	29'	05"	E
Rotterdam, Netherlands	4°	29'	46"	E
San Fernando, Spain	6°	12'		W
Santiago, Chile	70°	41'	00"	W
Singkawang, Indonesia (Island of Borneo)	108°	59'	41"	E
South Sumatra, Middle Meridian of, Indonesia	103°	33'		E
Stockholm, Sweden	18°	03'	30"	E
Sucre, Bolivia	65°	15'		W
Sydney, Australia	151°	12'	23"	E
Tenerife, Canary Islands	16°	35'		W
Tiranë, Albania	19°	46'	45"	E
Tokyo, Japan	139°	44'	40"	E
Washington, D.C., U.S.A.	77°	00'	34"	W

1. Exactly 20° west of Paris
2. South of Leningrad, seat of national observatory

B.3 *Guidelines to Determine the Scale of a Map*

TABLE 4.   Conversion to Greenwich Longitude[1]

For prime meridians with E longitudes:

To convert the E longitudes add the Greenwich differences.

To convert the W longitudes subtract from the Greenwich differences.

Example: Padang (100° 22′ 01″ E)

25°	31′	59″ E(P)	becomes 125°	54′		E(G)	(100°	22′	01″ +	25°	31′	59″)
112°	44′	E(P)	becomes 146°	53′	59″ W(G)	(112°	44′ + 100°	22′	01″)			
53°	42′	W(P)	becomes 46°	40′	01″ E(G)	(100°	22′	01″ −	53°	42′)		
121°	30′	W(P)	becomes 21°	07′	59″ W(G)	(121°	30′ − 100°	22′	01″)			

For prime meridians with W longitudes:

To convert the E longitudes subtract from the Greenwich differences.

To convert the W longitudes add the Greenwich differences.

Example: Ferro (17° 39′ 46″ W)

8°	22′	E(F) becomes	9°	17′	46″	W(G)	(17°	39′ 46″ −	8°	22′)
32°	14′	E(F) becomes	14°	34′	14″	E(G)	(32°	14′ − 17°	39′	46″)
63°	28′	W(F) becomes	81°	07′	46″	W(G)	(63°	28′ + 17°	39′	46″)
173°	25′	W(F) becomes 168°	55′	14″	E(G)	(173°	25′ + 17°	39′	46″)	

1. Slightly modified from table by C.B. Hagen.

TABLE 5.   Conversion from Centesimal to Sexagesimal Systems[1]

The sexagesimal division of the circle is now virtually universal in cartographic work. However, in the 18th century French scientists, using the metric system, devised the centesimal division of the circle. Today there exist large numbers of maps of France and its former colonial territories based on such a system. It can be quite confusing due to the relative closeness of the values.

The centesimal division of the circle is extremely simple. The entire circle is divided into 400 grads (a right angle is 90° in the sexagesimal system, 100 grads in the centesimal system). Each grad is in turn divided into 100 minutes and each minute into 100 seconds. The centesimal values can be expressed in regular decimal form or as minutes and seconds. The grad is shown as "G" and the centesimal minutes and seconds have the same marks as the sexagesimal ones, but with the slopes of the marks in the opposite direction.

Sexagesimal notation: 37° 23′ 12″
Centesimal notation: 48^G. 5734 or 48^G 57ˋ 34ˊˊ

The process of conversion is very simple.

It is known that 90° equals 100^G and that 60 sexagesimal minutes or seconds equal 100 centesimal minutes or seconds. Through a simple proportion multiply the centesimal values by 0.9 to obtain sexagesimal degrees and the remainders are multiplied by 60 to obtain sexagesimal minutes and seconds.

The latitude of downtown Saigon is 11^G. 9727 N or 11^G 97ˋ 27ˊˊ N.
Therefore: 11.9727 × .9 = 10.77543
.77543 × 60   = 46.5258
.5258   × 60   = 31.5480

A centesimal value of 11^G. 9727 or 11^G 97ˋ 27ˊˊ equals a sexagesimal value of 10° 46′ 32″ N.

1. Table by C.B. Hagen.

Appendix **C**

Date of Situation

C.1 The following procedures are suggested for the determination of the date of situation.

C.2 The dates appearing on the cartographic item provide a starting point from which to work. Examine the cartographic content which includes the following kinds of features:

1. airports, pipelines, railroads, roads, etc.
2. boundaries, both national and international
3. place names, e.g., countries, administrative divisions, cities
4. buildings and other man-made features on large-scale plans
5. names of persons or bodies responsible for producing the cartographic item.

Differences found when such features are compared and contrasted with similar features in the following available reference sources will provide an approximate date of situation.

1. *Cartactual.* — Budapest : Cartographia, 1966–
2. *Aktuelle JRO Landkarte.* — München : JRO Verlag
3. Encyclopedias– general, historical, national
4. Atlases– national atlases, world atlases, historical atlases, thematic atlases
5. Legal reference sources– constitutions, acts, etc.
6. Yearbooks and almanacs– world, e.g., *Statesman's Yearbook;* national
7. *Background Notes.* — U.S. Dept. of State
8. *International Boundary Study* and *Limits in the Seas.* Both published by U.S. Dept. of State, Office of the Geographer
9. Descriptive gazetteers, e.g., *Webster's Geographical Dictionary*
10. Authority files of bibliographic agencies– local, national
11. Histories– reliable histories of the area
12. Reference books– reliable ones relating to the area or subject
13. Calendars for the last several centuries
14. Archival collections
15. Bibliographies
16. Dictionaries of cartographers, surveyors, government officials, etc.

Series

D.1 Introduction

In cataloguing cartographic materials an understanding of the nature of map series is essential. A large proportion of any modern map collection is composed of various types of map series, and their correct identification is a necessary prerequisite for deciding how best to catalogue them. This appendix attempts to offer guidelines on what constitutes a map series and it gives some indication of the variety and identifiable characteristics of such series. Many map series, however, exhibit a number of similar features and it is often difficult to distinguish between them in any useful manner for the purposes of cataloguing.

D.2 General definition

A map series may be defined in general terms as:

> A number of related but physically separate and bibliographically distinct cartographic units intended by the producer(s) or issuing body(ies) to form a single group. For bibliographic treatment, the group is collectively identified by any commonly occurring unifying characteristic or combination of characteristics including a common designation (e.g., collective title, number, or a combination of both); sheet identification system (including successive or chronological numbering systems); scale; publisher; cartographic specifications; uniform format; etc.

This definition obviously bears a relationship to that for monographic series,[1] but it amplifies that definition to include certain identifying characteristics which apply specifically to cartographic materials, for example the sometimes significant unifying factor of scale. These factors often mean that a more sophisticated approach to the identification of map series is required than that normally applicable for monographic series.

D.3 Non-series maps

Map series should not be confused with multisheet single maps or maps composed of more than one sheet, or with map serials.

D.3A Multisheet single maps[2]

Properties and typical characteristics

The individual sheets of this sort of map should not be treated bibliographically as separate entities as they do not normally function as such, i.e., they are not intended for use independently by the map maker or producer.

The multisheet map is normally designed for use either by physical assembly or mounting as one entity. The resultant assembled map is a single map image and retains the primary characteristics of the ''single'' map printed on and complete in one sheet.

---

1. See Glossary for a definition of series.
2. A multisheet single map may be regarded as a multipart item (see *AACR2* Glossary).

For example, in such cases it is common for the individual sheets to be published or issued simultaneously, or within a very short time span, with the same publication and/or situation date for all the sheets by one and the same publisher, producer, or issuing agency. Similarly, updates, revisions, reissues, or new editions are done at the same time with the same date for all the sheets. (This is not necessarily the case for older cartographic works.) The sheets are not normally numerous and they are not usually available separately.

Often cartographic design features indicate that the group of map sheets is a multi-sheet map. Such features may include the occurrence of an overall single title for all sheets, or a single title and/or other text running across adjacent sheets to reveal itself fully only when all sheets are assembled, or borders designed in such a fashion that they only form a completed border when the map is assembled. The individual sheets may be numbered, lettered, or carry other designators. For the purpose of mounting the parts, a small index map may be provided in the border area, as an inset, or separately to show the location of the sheets for mounting purposes, e.g., Unesco's *Carte tectonique internationale de l'Afrique,* 1:5,000,000 in 9 sheets.

D.3B   Map serials

Map serials, although they occur infrequently, should be noted as exhibiting similar characteristics to those found in serials in general. The true cartographic serial is characterized by having individual number designations which are similar to those of periodials, i.e., having a serial title which appears unchanged from issue to issue and a numbering system which includes an issue number and date of issue.

D.4   Map series

D.4A   Types of map series and their identification

As a preliminary step in the identification of map series it is necessary to see whether the map in hand fulfills the requirements of the general definition of map series.

There are many types and characteristics of map series which are not necessarily self-evident. Of the different characteristics possible, only a selection may be possessed by any one series. Series vary from those in which area is the most important aspect, through thematic series in which subject is the main feature, to chronological series in which time assumes importance. To some extent these aspects are present in all map series, but they may assume varying degrees of importance.

D.4B   Contiguous area map series
Properties and characteristics

The chief criterion which distinguishes sheets of a contiguous area map series from non-series maps or from sheets of a multisheet single map is that the sheets exhibit the characteristics of single independent maps while maintaining a close relationship with all other sheets in the series, so that when publication is complete adjacent sheets can be trimmed and joined to form a larger map. The mapmaker's or producer's purpose in publishing the map series is also normally clear. The chief objective is to enable the individual sheets of the contiguous area map series to be used independently of each other, and most such series are so large that it would be impracticable to mount all the sheets for use as an entity, unlike the multisheet single map.

Most contiguous area map series are produced and published by a single authority, which may be a government body or a private or commercial firm. One publisher may, however, contribute towards a series in cooperation with others, particularly if the series is of worldwide or continental extent.

Contiguous area map series are usually relatively large cartographic works, sometimes involving as many as 60,000 sheets. The series as a whole usually has a series title and often has a series number; the individual sheets are normally named, or designated in some other way, i.e., they may bear numbers, letters, or a combination of both. The individual sheets are not necessarily published, reissued, or updated at the same time. The sheet designations do not in any way connote the chronology or sequence of publication of the sheets. At any given time any one sheet may have been issued a number of times, while another sheet in the same series may have been issued only once or may never have appeared. The lack of contiguous area sheet coverage extant at any given time does not therefore by itself constitute sufficient evidence that the series is not a contiguous area series. The publisher may by means of letters and/or numbers on each sheet indicate the juxtaposition of map sheets the one to another according to a pattern determined prior to publication; this will normally be evident from some sort of supplied graphic index.

Contiguous area map series are often long-term projects. The area to be covered, the sheet layout, the sheet designation system, the cartographic specifications, the revision and update procedures and their cycles, etc., are usually systematically laid out. The sheet characteristics are therefore frequently predictable, including uniformity of size, scale, projection, symbolization, colour schemes, and other technical cartographic specifications; most sheet boundaries are laid out systematically along a rectangular grid or geographical coordinate lines, i.e., on grid or graticule sheet lines. Nevertheless, individual sheets of contiguous area map series may tend with time to change considerably, due to the cumulative effect of minor changes in technical specifications. Thus currently issued sheets in a series that has been in progress for some time can exhibit considerable change as compared to sheets issued early on in the life of the series.

Contiguous area map series are most frequently general geographic maps (the topographic and planimetric series); some series are thematic, dealing with a specific subject such as soils, geology, land use, etc. Of all map series produced today, the contiguous area map series are the most common, the easiest to identify, and the best known.

>Australia 1:50 000 topographic survey
>
>Australia 1:100 000 topographic survey
>
>Australia 1:250 000 topographic survey
>
>Australia 1:250 000 geological series
>
>Canada 1:25 000
>
>Canada 1:50 000
>
>Canada 1:250 000
>
>The Canada land inventory series:

Land capability for wildlife – waterfowl 1:250 000

Land capability for wildlife – ungulates 1:250 000

Land capability for forestry 1:125 000 and 1:250 000

International map of the world 1:1 000 000

Karta mira 1:2 500 000

D.4C   Special or thematic map series
Properties and typical characteristics

In some map series the individual sheets may be used independently, and the concept of area is still quite important as a unifying factor; the prime characteristic is however the display of subject or thematic information. These map series are to be found normally as either (a) groups of maps each of a common area but displaying various themes or aspects of the same theme, e.g., geology, soils, rainfall for each month of the year, (b) a group of maps devoted to a common theme but not all of the same area, e.g., a group of maps indicating irrigation potential within a country, or (c) a group of maps, monographic in nature, which are either successively numbered or repetitively titled.

Sheets of subtype (a) are generally of uniform size, scale, and projection, etc., because it is most economical to produce them on a common base and superimpose the various subjects on that base. Sheets of subtype (b) and (c) on the other hand may be at different scales and different projections and show non-contiguous areas, though usually within a defined geographic entity.

In many cases the sheets of subtypes (a) and (b) will be published at about the same time and may not be available separately. As they are often the result of a specific project, the sheets are unlikely to undergo subsequent revision, but if revision does take place, it will almost certainly be applied to all sheets at the same time. The entire special map series will almost invariably be the work of one publisher who is likely to indicate the relationship between sheets by means of the designation, the sheet identification, or both.

East Africa 1:4,000,000. D.O.S. misc 299 A–G
(Sheets cover mineral resources; forest and game resources; vegetation; geology; soils; physical features.)

Emergency Measures Organization, urban analysis series : ₁Canada₁
– Vancouver (32 sheets)
– Toronto (32 sheets of each E & W)

Fitzroy region, Queensland : resources series

The individual items of subtype (c) exhibit a high degree of independence. Each item is monographic in nature, in that it has an independent publishing history, i.e., such items may be updated, revised, and reprinted independently of the other members of the unit, retaining the original assigned number or other designation, and can be used independently. There is no predictable frequency of publication of the items. The numbering system is systematic, but the logic behind it can often only be discovered by analyzing a whole group or by inquiry of the map producer in question. These groups are normally the work of one producer, publisher, or issuing body.

D.4D *Series*

Successively numbered map series

> Geological Survey of Canada "A" Series
> *(Caution: There are other prototype series imbedded in this map
> series viz., a contiguous area map series and an atlas.)*

> Miscellaneous geologic investigations (United States. Geological
> Survey) ; map I–

> Map series (Colorado Geological Survey)

> Geologic quadrangle map (United States. Geological Survey) ; G0–

Repetitively titled map series

> Bartholomew World Travel Map
> *(A "publisher's series")*

D.4D Chronological map series

Most chronological map series bear titles which may change from issue to issue, and
the sequential system may be based just on a date or an edition number rather than an
issue number as in the case of serials (see D.3B). A typical chronological map series
consists of one map which is reissued successively with the subject matter or information updated according to a predictable schedule.

> ₁Official highway maps₁

> ₁Some types of weather maps (monthly, weekly, daily)₁

D.5 Maps may also form part of a mixed media series where the grouping includes, for
example, volumes on a particular subject.

> Special publication (Montana. State Bureau of Mines and Geology)

D.6 Although the main types of map series have been listed here, mixed type series may
occasionally occur, so that the categories labelled above can only serve as guidelines to
the identification of map series. Maps in other formats may also on occasion be difficult
to distinguish from map series: for example, loose-leaf atlas sheets may be first published without a title page, contents list, or binding (see also Appendix F Geographical
atlases). The prime consideration for the cataloguer is to examine either all or as many
of the constitutent map sheets as possible for the design, bibliographic, and physical
characteristics of the map series and to consider what the map producer may have
intended.

Appendix **E**

Treatment of Map Series

The purpose of this appendix is to provide guidelines for the treatment of a collection of maps that has been identified as a series (see Appendix D for identification of map series). As some of the techniques presented here have not yet been rigorously tested, these guidelines are open to amendment.

E.0A   Selection of treatment method
The following factors influence the choice of cataloguing method for map series:

1. desirability of access to individual members of the series
2. economic factors, e.g., staff, budget, access to automated cataloguing systems
3. the number of members or potential members in a series and the holdings of the cataloguing agency
4. the relative strength of the relationships, i.e., the degree of independence or dependence among members of the series.

The following are general guidelines for determining the method to use for cataloguing a series:

1. If the number of sheets in a series is large and bibliographic access to individual members is not considered a priority, the most appropriate method to use is that of the description of the series as a whole, E.1 or E.2. These methods are particularly pertinent if there is no access to an automated cataloguing system.
    If, on the other hand, access to individual sheets is desirable, then multilevel cataloguing, E.4, (given an automated system) may be used, especially for contiguous area map series. A separate description, E.3, may be used for series that are monographic in nature.
2. If the number of sheets in a series is small, method E.3 or E.4 is appropriate if individual access is considered desirable and method E.1 or E.2 if individual access is not required.

Examples for some of the methods outlined below are given in 0K.

E.0B   Cataloguing methods for map series

Contents

E.4   Multilevel description (see also 13F)
Multilevel cataloguing provides complete identification of both
individual sheet and comprehensive whole.

E.1   Description of series as a whole

E.1A   All rules concerning descriptive cataloguing are applicable.
This method can be used in both manual and automated systems.

The method provides minimal bibliographic control through entry for the work as a
whole in the catalogue. It should be used in conjunction with a sheet index or a list to
provide access to the individual sheets of the series. Although bibliographic control of
the individual sheets is not provided by this method, it is the most economical for large
series when manual catalogue methods are employed.

Sheet index [1]

This provides access to sheets but not bibliographic access to individual parts, e.g.,
different editions of one sheet. Only the coverage held by the cataloguing agency is
marked on the sheet index.

List

This may be a brief listing organized by sheet numbers or names giving edition, date,
and any other information deemed essential by the cataloguing agency. This list provides some bibliographic information on individual sheets but no direct bibliographic
access to them. The listing may be on pages or sheets, or on cards.

E.1B   Specific applications

E.1B1   Title area. As many series are published over a long period of time, the
variation in title can be considerable.

   a)   When cataloguing a brand new series (i.e., a new publication) record the title
proper as found on the first sheets published if it is consistent and if it contains all
elements necessary for uniquely identifying the series.

   b)   If the title of a new or an older series varies or is incomplete, construct a title
proper as instructed in the application to 1B7. Record the title variations in a note
(7B4). Added entries may be made at the discretion of the cataloguing agency.
Disregard minor variations when recording variations in the title proper.

E.1B2   Statement of responsibility

   a)   If the statement of responsibility is a corporate body and the form of its name
changes, record the earliest form of the name used in the series. Subsequent
forms of the name are recorded in a note and added entries made for them.

   b)   Some series are produced cooperatively (mainly world series), and each partici-
pant is responsible for producing the maps of a specified area, e.g., the International map of the world 1:1 000 000 and Karta mira 1:2 500 000.

---

1. A record may be created for a printed sheet index.

If there are three or fewer cooperating bodies record them in the statement of responsibility. If there are four or more, choose one corporate body and record it according to the instructions in 1F5. Record the other responsible bodies in a note (7B6) and make added entries for them.

E.1B3 Mathematical data area. Projection statement

List up to three projections in the projection statement. Any explanations may be made in the note area (7B8). If there are four or more list the three most commonly occurring projections. The remainder may be recorded in a note.

E.1B4 Publication, distribution, etc., area

a) If the name of the publisher changes over the course of the publishing history, record the earliest form of the name occurring in the series. Record subsequent forms of the name in a note and make added entries for them.

b) For series published cooperatively where each body is responsible for only a part of the whole, all places and publishers, etc., may be recorded if there are three or fewer (see 4D6). If there are four or more, record ₍Various places : various publishers₎.[2]

E.2 DESCRIPTION OF SERIES AS A WHOLE WITH CONTENTS NOTE (see 13C)

The descriptive rules for cataloguing and rule 7B18 are applicable.

This method can be used in both manual and automated systems.

The contents note is used only when a short citation of the sheets or volumes of a series or atlas is needed and the number of sheets or volumes is limited. In theory, however, a full citation for each sheet or volume may be made and displayed.

Access points may be provided for both the description of the work as a whole and its contents. Name-title added entries give direct access to individual sheets or volumes.

E.3 SEPARATE DESCRIPTION WITH SERIES STATEMENT (see 13B)

Rules for description are applicable (see area 6 for series statement).

This method can be used in both manual and automated systems.

When this method is used, each individual sheet of the series may have added entries for author, title, and subject. Series added entries give access to all sheets of the series under one and the same heading.

E.4 MULTILEVEL DESCRIPTION (see 13F)

The method for multilevel cataloguing is in rule 13F Application. These descriptive rules provide guidance for recording information at the first and subsequent levels.

The method is recommended for use only in automated systems.

E.4A The basic principle of the multilevel cataloguing method is to record at the first level only information relating to the series as a whole. Information recorded at the second level pertains only to the individual sheets. This property of non-repetitiveness of bibliographic data makes the multilevel cataloguing method economical and attractive to

---

2. This problem is unresolved in *AACR2* Part I, so this method has been introduced to provide a reasonable solution.

apply to series with a large number of sheets where individual description and access to the parts is required. It should be noted that access points can be attached to each second (and subsequent) level description as well as to the first level.

Description layout and processing are often problemmatical when the multilevel cataloguing method is used. Second-level descriptions are incomplete and always need to be displayed with the first-level catalogue record in order to fully identify the sheet within the series. For instance, a series with a uniform scale throughout will have this scale recorded only at the first level.

Another problem is the instability of first-level records for current series until such time as those series close or discontinue. However, the rules are designed to provide first-level records that are as stable as possible under the circumstances.

Because the multilevel method is complex and labour intensive, it is not recommended for use in manual catalogues. An automated system can provide a machine link between the two (or more) levels through, for example, the use of control numbers. In such systems this method is both economical and efficient.

E.4B Specific applications

E.4B1 Title proper. At the first level, record the title proper according to E.1B1. Record sheet titles at the second level. Do not indicate the omission of sheet titles at the first level by the mark of omission.

If a sheet has both a sheet title and a sheet number, record the sheet number first and separate the title from it by colon, space.

E.4B2 Statement of responsibility. At the first level, record the statement(s) of responsibility according to E.1B2. Responsibility for specific parts is recorded at subsequent levels.

E.4B3 Edition area. If a sheet has more than one edition statement, each assigned by a different agency (e.g., military and civilian agencies), record all of them in a second-level description. Record first the edition statement pertinent to the first-level description. The remaining edition statements are recorded as subsequent edition statements. Add the appropriate statement of responsibility to each edition statement.

> — Ed. 2-GSGS / Ministry of Defence. ₍Ed.₎, A ₍bar, star₎ / Ordnance
> Survey
> *(For the catalogue entry of the military edition)*

E.4B4 Publication, distribution, etc., area. At the first level, record the publisher, etc., according to E.1B4. The publisher for specific sheets is recorded at subsequent levels.

E.4B5 Physical description area. Record physical details relevant to all the sheets at the first level. Variations for specific sheets are recorded at subsequent levels.

E.4B6 Note area. Notes are recorded following the level to which they pertain, i.e., notes relating to the series as a whole are recorded at the first level, while those relating to one sheet, are recorded at the second level.

Appendix **F**

Geographical Atlases

Contents:

F.1   INTRODUCTION

An atlas is a systematically arranged collection of maps or charts intended to be shelved like a volume, either flat or vertically. Historically, many geographical atlases originated as collections of maps or charts, produced for a particular purpose, and assembled in the form of volumes, portfolios, or cases meant to be shelved.

Atlases share the characteristics of maps and books, having the content of maps and often the format of books. Like maps, most geographical atlases consist of graphical representations of the earth's surface; some include aerial photographs, photomaps, and remote-sensing imagery.

F.2   DISTINGUISHING CRITERIA

An atlas in book form is distinguished from other books by its emphasis on maps: any textual material is secondary or complementary and serves mainly to support and explain the cartographic content.

An atlas is distinguished from other collections of maps primarily by the publisher's or compiler's apparent intent that the work be used like a book and shelved in the form as issued.

F.3   FORMAT

Atlases are published in a variety of formats. For example, they may be bound or loose-leaf, flat or folded, in cases, boxes, portfolios, or folders. Sizes range from miniature and small pocket size through quarto, folio, and even larger sizes.

The cataloguing record describes the atlas as received, and atlases should be preserved and stored as closely as possible to the original format. On a selective basis, for the purpose of preservation, the format may be altered, e.g., folded maps may be unfolded; covers of parts isolated from contents (but kept with them); highly unstable material isolated; leaves removed from original bindings for deacidification and encapsulation prior to being rebound.

F.4   SCOPE

Like other cartographic materials, atlases show, in graphic form, the geographical distribution of information. Their content may be general, regional, thematic, or special purpose (e.g., braille). They may cover areas as large as the known universe, or as small as a part of a school campus. They may include only bodies of water or land areas, or they may include both.

F.5 BIBLIOGRAPHIC ASPECTS

Atlases may or may not contain a title page and other expected parts such as tables of contents, preface, introduction, appendices, and indexes. Sometimes explanatory texts are issued under separate cover.

Atlases often include an index map to the collection of map sheets which forms the body of the atlas. They frequently have indexes to place names, streets, landowners, tourist sites, etc., and they may include directories.

Atlases usually have legends of map symbols, and frequently contain graphs, tables, gazetteers, glossaries, bibliographies, description and travel information, and even biographies and histories. Any textual material is secondary.

The key words in an atlas title may be "atlas of . . . ₁geographic area₁"; "map of . . . "; "guidebook for motoring in . . . "; "guide to . . . "; etc. In any case, the item is treated as an atlas if it is principally a collection of maps issued in a form intended to be shelved.

The atlas spine, portfolio, or container often bears one or more elements of bibliographic data: title, author, publisher, distributor, edition, or date of publication.

Appendix **G**

Examples

This appendix provides examples of full bibliographic descriptions illustrating rules and applications. It is organized into two parts, the concordance table and the examples themselves.

The concordance refers from the rule number (or its application) to the examples illustrating that rule or application. The examples are organized by the name of the country of the contributing agency, each one being numbered consecutively within a block of 100 numbers set aside for each country. For example, the United States has the 400 block, which has been subdivided into sections for maps (400–449) and atlases (450–499). Each example is followed by the rule number or numbers that its contents illustrate. The number of examples for any one rule has been limited to include a variety of, but not necessarily all, possibilities.

Concordance for Examples

RULE NUMBER		EXAMPLE NUMBER	RULE NUMBER		EXAMPLE NUMBER
0.4		3, 21, 39, 57	1B3		210
0B4		465	1B4	Application	311, 313
0B7		463	1B5		6, 431
0F1		40, 50	1B7		16, 33, 208, 463
1	Application	35	1B7	Application 2	211
1B1a		20, 23, 38, 205, 310, 470, 478	1B8a		45, 300
			1B8a	Application	26
1B1a	Application	5, 466	1B8b	Application 1	10, 45, 200, 204, 412, 427
1B1c	Application	22			
1B2		15, 24, 48, 53, 305, 473	1B8b	Application 2	201

RULE NUMBER		EXAMPLE NUMBER	RULE NUMBER		EXAMPLE NUMBER
1B9		22	4C1		62, 431
1B10		5, 432	4C3		7, 21, 409, 429, 451,
1B12		465			464
1B14		305	4C5		37, 58
1B15		306	4C6		8, 13, 19, 52, 210, 470,
1D1		45, 54, 451			471
1D2		453	4C7		12, 305
1E1		10, 56, 305, 409, 470	4D1		423
1E2		460	4D3		214, 308, 422
1E4		305	4D4		403, 451, 461
1E6		9, 11, 47, 54, 210, 307,	4D5		210, 216, 400, 424, 425,
		407			429, 472, 475, 476
1F1		25, 205, 212, 304, 305,	4D7		13, 19, 49, 470, 471
		308, 423, 424, 427,	4E1		68, 408
		451, 459	4F2		313, 475
1F2		9, 31, 417, 419	4F5		301
1F3		20, 44, 55, 63, 414, 472,	4F6		62, 69, 416, 422, 426,
		475			427, 452, 469
1F4		41, 49, 454, 476	4F7		23, 27, 30, 35a, 208,
1F5		426, 458			210, 217, 300, 429,
1F6		6, 23, 50, 55, 450, 456,			455, 470
		459, 469, 475	4F9		16, 29, 205
1F7		49, 50	4F10		207, 311, 312
1F8		16	4G1		66, 471
1F10		22	4G4		32, 59, 71, 415, 418
1F11		4, 6, 46	5	Application	35
1F12		2, 28, 41, 60	5B1a	Application	55, 61, 64, 67, 72, 202,
1F13		18, 210			302, 406, 411, 430
1F17		208, 307	5B2a		33
1G3		65, 421	5B2b		7, 25, 48, 200, 204, 407,
2B1		42, 50, 66, 309, 403,			421
		432, 454, 472, 477	5B5		454, 467, 472
2C1		70, 410	5B6		22, 472, 477
2D1		301	5B7		473
2D2		474	5B12		457
3	Application	35	5B13		478
3A2		457	5B14		478
3B	Application	421	5B20		460
3B1a		20, 416, 472	5B20	Application	453
3B1a	Application 2	453	5B21		470
3B1b		427	5B22		451
3B1d		7, 13, 208, 305, 432	5B26		462, 463, 464
3B1e		57, 61, 202	5B28		213, 214
3B2		6, 10, 34, 212, 412, 415,	5B30		57
		425, 427, 455, 468,	5C1	Application	29, 42, 203, 208, 402,
		473, 478			421, 430
3B2	Application	207	5C2		471
3B3		37	5C2a		22, 470, 475
3B4		7, 35a, 219, 468	5C2c		458
3B5		25, 204	5C2d		468
3B6		470	5C2f		474
3B7		411, 430	5C3		8, 453
3B8		64, 405	5C3	Application	14, 465
3C1		37, 218, 412, 420, 429	5C4		43, 50, 64
3C2		8, 59, 63, 404	5C5		53, 72, 406
3D1b		429	5D1a		14, 205
3D1b	Application	12, 49, 300, 412	5D1b		14, 209
	Application 2	53	5D1b	Application	413
	Application 7	2	5D1c		53
3D2		302, 310	5D1e		425
4A2		11	5D1f		33
4B6		15	5D1g		28, 48

1 Canadian Examples

Canadian Examples

1.

   Atlantic Provinces. -- Scale [1:1 500 000]
(W 72°15'--W 59°50'/N 48°30'--N 43°30'). --
Toronto, Ont. : Creative Map Sales Ltd., c1976.
   1 map : col. ; 42 x 63 cm, folded to
22 x 10 cm.
   Copyright, Creative Sales Corporation, 1972-
1976.
   Inset: Quebec north shore.   Scale
[1:3 294 720].   5 x 11 cm.
   On verso: Newfoundland.   Scale [1:1 900 800].
33 x 28 cm — Cape Breton Highlands National
Park. 17 x 28 cm — Cape Breton Island area.
6 x 10 cm — Maritime Provinces. 8 x 28 cm —
Halifax. 8 x 14 cm — St. John's. 7 x 14 cm.
   Includes text and ill., time and distance
chart, and partial list of location of cities
and towns.

(7B9; 7B18b Application 5, 6, 7)

2.

   Bugaboo Glacier, British Columbia, Canada /
survey, photography, photogrammetry and
compilation by the Inland Waters Branch,
Department of Energy, Mines and Resources. --
Scale 1:2 500 (W 116°45'/N 50°40'). -- Ottawa
[Ont.] : Inland Waters Branch, 1969 (Surveys
and Mapping Branch).
   1 map : col. ; 48 x 80 cm. -- (Glacier map
series no. 3B ; sheet no. 5)
   "Rectangular Co-ordinate System chosen
arbitrarily ..."
   Accompanies report: Glacier surveys in
British Columbia, 1968 : vol. 2 metric system /
I.A. Reid and J. Shastal. Ottawa, Ont. :
Inland Waters Branch, [1970]. (Report series ;
no. 10).

Inset: Key map : [showing location of seven glaciers in Alberta and British Columbia]. 9 x 10 cm.

(1F12; 3D1b Application 7; 6G1; 7B8; 7B11)

3.
Canada. -- Scale [ca. 1:14 709 000] (W 141°-- W 50°/N 75°--N 41°). -- Toronto [Ont.] : Regal Stationary Co., [196-?].
1 jigsaw puzzle (208 pieces) : col. ; 26 x 43 cm, in box 26 x 27 x 6 cm.

(0.4; AACR 2 10.5B2, 10.5C)

4.
Canada census divisions and subdivisions, 1971 / compiled and produced by National Geographical Mapping Division = Canada divisions et subdivisions de recensement, 1971 / dressée et établie par la Division des cartes géographiques nationales. -- Scale 1:7 500 000. -- Ottawa [Ont.] : National Geographical Mapping Division, 1978.
1 map : col. ; 73 x 73 cm. -- (The national atlas of Canada ; 5th ed.)
Insets: Windsor-Quebec. Scale 1:2 000 000. 35 x 93 cm — Vancouver-Victoria. Scale 1:2 000 000. 9 x 9 cm — Winnipeg. Scale 1:1 000 000. 8 x 9 cm — Island of Newfoundland = Ile de Terre-Neuve. Scale 1:5 000 000. 12 x 12 cm.
"Index to census divisions and subdivisions". Map no.: MCR 4000.

(1F11)

5.
Canada, photo mosaic series 50, 1:50 000 / Army Survey Establishment. -- Scale 1:50 000 ; transverse Mercator proj. -- [Ottawa, Ont.] : Army Survey Establishment, [1963?]-1964.
28 remote sensing images ; 56 x 65 cm and 56 x 32 cm.
"Universal transverse Mercator grid zone 12." North American datum 1927.
Military series no. A 042.
Series is a special purpose military training area series.

(1B1a Application; 1B10; 5D1k; 7B8; 7B12)

6.
Canada, showing location of medical services branch facilities / produced by the Surveys and Mapping Branch, Department of Energy, Mines and Resources ; information compiled by the Department of National Health and Welfare (Medical Services Branch) = Canada et l'emplacement des installations de la Direction générale des services médicaux / établie par la Direction des levés et de la cartographie, Ministère de l'énergie, des mines et des

ressources ; les renseignements ont été rassemblés par le Ministère de la santé et du bien-être social (Direction generale des services medicaux). -- Scale 1:6 336 000. 1 in. to 100 miles (W 170°--W 50°/N 80°--N 40°). -- Ottawa [Ont.] : Surveys and Mapping Branch, 1978.
1 map : col. ; 73 x 93 cm.
Inset: [Vancouver Island and part of south western British Columbia]. Scale 1:2 000 000. 1 in. equals approx. 32 miles (W 128°30'-- W 119°30'/N 50°20'--N 48°00'). 25 x 33 cm.

(1B5; 1F6; 1F11; 3B2)

(Note: *Verbal scale statement on map is "one inch to 100 miles." "Canada" appears only once.*)

Canada's Grand Trunk and Grand Trunk Pacific Railway system. -- Scale [ca. 1:5 200 000 and ca. 1:9 500 000]. -- Chicago [Ill.] : Poole Bros., 1910.
2 maps on 1 sheet : col. ; 23 x 61 cm and 39 x 61 cm.
Title from verso.
Provenance: Sir Alfred Waldron Smithers, 1850-1924 [Papers].
Components: Eastern section : Canada's Grand Trunk Railway system and connections [and] Canada's Grand Trunk Pacific Railway — Western section : Canada's Grand Trunk Railway system - connections west of Chicago [and] Canada's Grand Trunk Pacific line west of Fort William.
Ill. on verso.

(3B1d; 3B4; 4C3; 5B2b)

8.
Carte électorale du Québec, 1973 / préparée et publiée par le Service de la cartographie. -- Scale 1:1 250 000 ; conformal conical proj. with two standard parallels 46° and 60° (W 81°--W 56°/N 53°--N 45°). -- [Québec? Québec] : Service de la cartographie, 1975.
1 map : col. ; 87 x 140 cm.
"3e impression janvier 1975".
Insets: Partie nord du Québec. 18 x 24 cm — Québec et environs. Scale [1:56 300]. 25 x 29 cm — Montréal et environs. Scale [1:112 640]. 25 x 45 cm — Taillon. Scale [1:56 300]. 14 x 10 cm — Laprairie. Scale [1:56 300]. 14 x 10 cm — Laporte. Scale [1:362 000]. 14 x 10 cm — Hull. Scale [1:126 720]. 10 x 9 cm — Trois Rivières. Scale [1:144 800]. 14 x 9 cm — Sherbrooke. Scale [1:68 200]. 10 x 9 cm.
Lists electoral districts.

(3C2; 4C6; 5C3; 7B9)

9.

Catégorie économique des fermes par comté en 1970 : [Québec]. -- Scale [1:1 400 000]. -- [Québec, Québec] : C.E.G.E.P. de Limoilou, 1978.
1 map : col. ; 40 x 54 cm.
Map design by Réal Allie, according to a letter from the C.E.G.E.P.
"Source: Statistique Canada."
Insets: [Zones périphériques]. Scale [1:12 300 000]. 7 x 13 cm.

(1E6; 1F2; 7B6)

10.

Census metropolitan area Calgary, Alberta : average household income for 1971 / the Financial post survey of markets computer income maps produced in association with Lanpar Limited. -- Scale [1:55 000] not "1:50 000". -- [Toronto, Ont.] : Financial post, [1976?].
1 map ; 45 x 37 cm.
Title and statement of responsibility from verso.
Title on recto: Calgary.
Computer line print map.
Description of market area on verso.

(1B8b Application 1; 1E1; 3B2; 7B10a)

11.

City of Thunder Bay : [showing polling subdivisions and ward boundaries]. -- Scale [1:48 888] (W 89°26'--W 89°09'/N 48°31'-- N 48 17'). -- [Thunder Bay, Ont.] : [City of Thunder Bay] Planning [Dept.], 1976.
1 map : 58 x 48 cm.

(1E6; 4A2)

12.

Colton's New Brunswick, Nova Scotia, Prince Edward Id. & Cape Breton Id. -- Scale [ca. 1:1 548 000] (W 71°50'--W 59°00'/N 49°30'-- N 43°15'). -- New York [N.Y.] (172 William St.) : G.W. & C.B. Colton, [1862?].
1 map : hand col. ; 42 x 62 cm.
Variant of sheet 21-22 of: Colton's general atlas, containing one hundred and eighty steel plate maps and plans, on one hundred and nineteen imperial folio sheets / drawn by G. Woolworth Colton ; letter-press descriptions, geographical, statistical, and historical by Richard Swainson Fisher. New York [N.Y.] : G.W. & C.B. Colton & Co. ; London [England] : Bacon & Co., 1865. 46 x 40 cm. A copy of the atlas is in the National Map Collection.
Meridian of Washington.
Latest date mentioned on the verso text is 1862.
"Entered according to Act of Congress in the year 1855 by J.H. Colton & Co. in the Clerks Office of the District Court of the United States for the Southern District of New York."
On verso: descriptive text on Prince Edward

Island, Newfoundland, and Nova Scotia.
Decorative border.

(3D1b Application; 4C7; 7B7a; 7B8)

13.

Counties of Prescott and Russell, and Stormont, Dundas, and Glengarry. -- Scale [ca. 1:277 740]. -- [S.1. : s.n., 197-?].
1 map : photocopy ; 36 x 48 cm.
Copied from: Canada, past, present and future, being a historical, geographical, geological and statistical account of Canada West : containing ten county maps and one general map of the province, compiled expressly for the work / by W.H. Smith. Toronto, Ont. : Thomas Maclear, [1852]. Vol. 2.
The original published in Toronto, Ont. by Thomas MacLear.

(Note: *This is a photocopy of an original map that is bound with the cited monograph.*)

(3B1d; 4C6; 4D7; 7B7a; 11F)

14.

County map of Nova Scotia, New Brunswick, Cape Breton Id. and Pr. Edward's Id. -- Scale [ca. 1:3 000 000] (W 68°20'--W 59°40'/N 50°35'-- N 42°50'). -- [Philadelphia, Pa.] : S. Augustus Mitchell, 1877.
1 map : hand col. ; 29 x 23 cm, on sheet 39 x 31 cm.
From: Mitchell's new general atlas, containing maps of the various countries of the world, plans of cities, etc. ... Together with valuable statistical tables. Also a list of post-offices of the United States and territories, and census of 1860 and 1870. Philadelphia, Penn. : S.A. Mitchell, 1878.
"Entered according to Act of Congress in the year 1877 by S. Augustus Mitchell in the Office of the Librarian of Congress."
Atlas listed in: A list of geographical atlases in the Library of Congress / P.L. Phillips, v. 1, pt. 1, title 888, p. 510.
Inset: City and harbour of Halifax. Scale [ca. 1:156 000]. 7 x 6 cm.
On verso: Map of Quebec in counties. Scale [ca. 1:2 027 520]. 23 x 30 cm. Inset: Environs of Montreal. 8 x 10 cm. Plate no. 8.
Plate no. 7.

(5C3 Application; 5D1a, b; 7B7a; 7B15)

(Note: *A separate description can be made for the map on verso, with linking notes.*)

15.

The Davenport map of British Columbia and adjacent territories, showing main river & transportation systems. -- Scale

[ca. 1:1 500 000] (W 139°--W 113°/N 60°--N 56°).
-- Toronto [Ont.] : For sale by Angus Mack &
Company, c1929.
  1 map : col. ; 132 x 117 cm, in folder
33 x 21 cm.
  Copyrighted by Joseph B. Davenport.
  Original imprint, covered by label, reads:
Compiled and published by the Island Blue Print &
Map Co., Sayward Building, Victoria, B.C.
  Inset: Continuation north westward at same
scale, 40 x 37 cm.

(1B2; 4B6; 7B9)

16.
  [Dog Creek Indian Reserve 46, Manitoba] /
[signed] H. Martineau. -- Scale [1:63 360]. 50
chains to 1 in. -- [ca. 1888].
  1 map : ms. ; 23 x 23 cm.
  Legend names Indians and provides a key to
their residences and includes their religious
denomination.
  Ink drawing.
  Stamp: "Manitowah agency".

(1B7; 1F8; 4F9; 7B20a)

17.
  Excelsior series map of United States and
Alaska / edited, engraved and published by the
Geographical Publishing Co. -- Scale
[1:4 752 000]. 75 miles to 1 in. (W 127°--W 65°/
N 50°--N 24°). -- Chicago, Ill. : Geographical
Publishing Co., [ca. 1940].
  1 map : col. ; 60 x 89 cm.
  "Largest cities of the world showing the latest
population. Population from the latest official
census and latest estimates."
  Provenance: Vasile Avramenko, 1894-
[Papers].
  Inset: Alaska. Scale [1:14 952 960]. 236
miles to 1 in. (W 180°--W 120°/N 70°--N 52°).
15 x 18 cm, with inset showing continuation of
Aleutian Islands, 3 x 9 cm.
  On verso: Portraits of the United States
presidents and state flags.

(7B18b Application 5, 7)

18.
  Foster's cyclists' road map of eastern
Ontario. -- Scale [1:380 160]. 6 miles to 1 in.
(W 79°30'--W 73°30'/N 45°39'--N 43°37'). --
Toronto Ont. : J.G. Foster & Co., 1900.
  1 map : both sides ; 66 x 121 cm, on sheet
48 x 72 cm.
  "Entered according to Act of Parliament of
Canada, in the year 1900, by J.G. Foster & Co.,
at the Department of Agriculture."
  On verso: Westward continuation of main map,
at same scale. 34 x 66 cm.
  Folder: Inside cover includes extracts from
provincial traffic legislation and quotes traffic
by-laws of the City of Toronto.
  Includes place index listing population totals
and legend.

NMC copy has been conserved. Map has been
separated from folder (18 x 12 cm) for flat
storage.

(1F13; 5D1j)

19.
  Geological map of northern Israel / prepared
by L. Picard. -- Scale 1:100 000 (E 34°47'--
E 35°41'/N 33°27'--N 32°23'). -- [S.l. : s.n.,
195-].
  1 map : col. ; 121 x 84 cm.
  In Hebrew and English
  Provenance: Bernard M. Bloomfield, 1904-
[Papers].
  Working document. Parts of two sheets of
Israeli geological survey and four legends
mounted on linen. Title block printed by hand
and mounted on linen.

(4C6; 4D7; 7B2; 7B7b; 7B10a)

20.
  The gold and coal fields of Alaska, together
with the principal steamer routes and trails /
Department of the Interior, U.S. Geological
Survey. -- Scale 1:3 600 000. 57 miles to 1 in.
(W 170°--W 128°/N 72°--N 54°). -- [Washington,
D.C. : U.S. Geological Survey], 1898.
  1 map : col. ; 59 x 71 cm, folded to
24 x 15 cm.
  "The base map is a copy of a part of chart T
of the U.S. Coast and Geodetic Survey."
  Accompanied by text: Map of Alaska showing
known gold-bearing rocks with descriptive text,
containing sketches of the geography, geology,
and gold deposits, and routes to the gold
fields. 44 p. 24 cm. (G4371.H1 1898.U5 c.2)
  Insets: Trails from Tide Water to the
headwaters of the Yukon River, Alaska. Scale
1:1 447 000 (W 136°--W 132°/N 61°--N 58°).
22 x 16 cm -- The Klondike gold region, Canada.
Scale 1:1 447 000 (W 142°00'--W 138°00'/N 65°00'
--N 62°30'). 19 x 16 cm.

(1B1a; 1F3; 3B1a; 7B11; 7B18b Application 5)

21.
  Industrial life of Canada = La vie
industrielle au Canada. -- Scale
[ca. 1:17 160 000] (W 180°--W 20°/N 80°--N 42°).
-- Bournemouth, England : G.J. Hayter & Co.,
[196-?].
  1 jigsaw puzzle (125 pieces) : col., plywood ;
25 x 39 cm, in box 27 x 42 x 3 cm. -- (Victory
geographical puzzles ; 7485)

(0.4; 4C3; AACR 2 10.5B2, 10.5C)

22.

Integrated development study Vieux Port/Bassin
Louise, Quebec = Etude sur un développement
intégré Vieux Port/Bassin Louise, Québec.
Propriétés fédérales à Québec secteurs Vieux
Port/Bassin Louise : [volume] 3, les cartes /
cartes préparées par la Direction de
l'aménagement urbain et Ministère des travaux
publics. -- Scales differ. -- [Ottawa, Ont.] :
Direction de l'aménagement urbain et Ministère
des travaux publics, 1976.
   1 atlas ([3], 10 leaves) : 10 maps ; 43 cm.
   Report consists of three volumes, volume 1 in
English, volumes 2 & 3 in French. Subtitle of
volume 1: Conceptual programme. Subtitle of
volume 2: Analyse du marché.
   Contents: 1. Québec métropolitain. Scale
1:75 000 — 2. Québec centre, haute ville et
basse ville. Scale [ca. 1:12 000] — 3.
Secteurs de développement. Scale [ca. 1:4 800]
— 4. Autres programmes et projets sur la zone.
Scale [ca. 1:4 800] — 5. Circulation. Scale
[ca. 1:4 800] — 6. Propriétaires fonciers.
Scale [ca. 1:4 800] — 7. Propriétés fédérales.
Scale [1:4 800] — 8. Contraintes. Scale
[ca. 1:4 800] — 9. Facteurs prédominants.
Scale [ca. 1:4 800] — 10. Proposition
d'utilisation du sol. Scale [1:4 800].
   "C.C. no. 31. Confidential".

(1B1c Application; 1B9; 1F10; 5B6; 5C2a; 5D2;
7B18b Application 4)

23.

Iron ore deposits at Bessemer, lot 1, con.
VII, 2, 3, 4 and 5, con. VI, township of Mayo,
Hastings County, Ontario / H.E. Baine ; L.H.S.
Pereira. -- Scale 1:2 400. 200 ft. to 1 in. --
[Ottawa, Ont.] : Dept. of Mines, Mines Branch,
[ca. 1912].
   1 map : col. ; 36 x 89 cm.
   Geology by E. Lindeman, 1911.
   In appendix to volume I of accompanying
report: Iron ore occurrences in Canada /
compiled by E. Lindeman and L.L. Bolton.
Ottawa, Ont. : Govt. Printing Bureau, 1917.
Vol. I: Descriptions of principal iron ore
mines. 71 p. Report no. 217.
   Map no. 191A.

(1B1a; 1F6; 4F7; 7B6; 7B11; 7B19)

24.

Lloyd's new map of Canada showing railways in
colours, air mail routes, natural resources,
proposed Trans-Canada automobile highway,
Quebec-Labrador boundary. -- Scale [1:3 801 600].
60 miles to 1 in. (W 150°--W 40°/N 73°--N 40°).
-- Toronto [Ont.] : Angus Mack Co., [ca. 1928].
   1 map : col. ; 95 x 150 cm, in folder
35 x 21 cm.
   Includes index to places, with population
statistics, and table of population by provinces.
   Inset: Map of the Dominion of Canada showing
air mail and proposed routes [in red], proposed

Trans Canada automobile highway [in green].
26 x 38 cm.

(1B2; 7B1a)

25.

A look at Canada / designed and produced for
the Department of the Secretary of State, by
the Surveys and Mapping Branch, Department of
Energy, Mines and Resources. -- Scales differ
(W 142°--W 50°/N 83°--N 41°). -- Ottawa [Ont.] :
Surveys and Mapping Branch, c1977.
   6 maps on 1 sheet : col. ; 34 x 35 cm or
smaller, on sheet 56 x 87 cm.
   Also available in French.
   Components: Population. Scale 1:20 000 000
— Economic resources. Scale 1:15 840 000 —
Climatic regions. Scale 1:35 000 000 — Natural
vegetation. Scale 1:35 000 000 — Relief. Scale
1:20 000 000 — Political. Scale 1:15 840 000.
   Inset: [Location of Canada].
   Text and ill. on verso.

(1F1; 3B5; 5B2b; 7B18b Application 4, 5, 6)

26.

Major oil and gas fields and pipelines in
Canada = Les principaux champs pétrolifères et
gazifères et les principaux pipelines au Canada.
-- Scale [ca. 1:15 600 000] (W 150°--W 50°/
N 80°--N 40°). -- Ottawa [Ont.] : Energy Policy
Sector, 1978 (Surveys and Mapping Branch, Dept.
of Energy, Mines and Resources).
   2 maps : back to back, col. ; 32 x 37 cm each.
   English and French versions of same map.
   Inset: [Location of major oil and gas fields,
western Canada]. Scale [ca. 1:6 600 000].
22 x 17 cm.
   Includes index of major oil and gas fields, and
legend.

(1B8a Application; 7B2)

27.

Map of Alaska & surroundings, showing the
Klondike Gold Fields, and routes to the mining
camp. -- Scale [ca. 1:4 700 000] (W 170°--W 130°/
N 68°--N 54°). -- [S.l. : s.n., 19-?].
   1 map : col. ; 40 x 53 cm.
   Inset: Map of Klondike region. 9 x 11 cm.

(4F7)

28.

Map of Canada West or Upper Canada / compiled
from government plans, original documents and
personal observation by William H. Smith. --
Scale [ca. 1:448 230] (W 84°--W 74°/N 47°--N 41°).
-- Toronto : Thomas Maclear, 1852.
   1 map : hand col. ; 43 x 57 cm, on sheet
74 x 95 cm.
   Additional data pasted on sheet: View of
Kingston, C.W. — View of Cobourg, C.W. — View
of London, C.W. — Port Hope — Queenston
suspension bridge — Brockville, C.W. —

"Statistics Canada, according to the census of 1852".
   Accompanies: Canada : past, present and future, being a historical, geographical, geological and statistical account of Canada West : containing ten county maps and one general map of the province, compiled expressly for the work / by W.H. Smith. Toronto : Thomas Maclear, [1852].

(1F12; 5D1g; 7B10a)

29.
   Map of land to be drained by the Cavanagh ditch, Kitley / Walter Beatty. -- Scale [ca. 1:4 752]. 6 chains to 1 in. -- 1904.
      1 map : ms. ; 62 x 37 cm.
      Ink drawing.

(4F9; 5C1 Application)

30.
   Map of Temiskaming and Northern Ontario Railway. -- Scale [ca. 1:620 000] (W 83°30'-- W 78°30'/N 50°00'--N 45°45'). -- [S.l. : s.n., 190-].
      1 map : col. ; 76 x 58 cm.
      Manuscript additions: Tp. survey plat. Sept. 29/06.

(4F7; 7B20a)

31.
   Map of the Dominion of Canada. -- Scale 1:6 336 000. 100 miles to 1 in. (W 142°--W 46°/ N 70°--N 40°). -- [Ottawa, Ont.]. : Dept. of the Interior, 1916.
      1 map : col. ; 49 x 88 cm.
      Shows railways. Lists 1915 statistics for railway construction, railway mileage operated, and railway mileage by province.
      "J.E. Chalifour, Chief Geographer."
      Copy 2: Has ms. addition of Canadian Northern Railway System. Label attached with key to colours used. Acc. no. 78902/4812.

(1F2; 7B1a; 7B6; 7B20a)

32.
   Map shewing the general northern limits of the principal forest trees of the Dominion of Canada / by Robert Bell. -- Scale [1:5 068 800]. 18 miles to 1 in. (W 138°--W 50°/N 70°--N 40°). -- [Ottawa, Ont.] : Geological Survey of Canada, 1881 (Montréal [Quebec] : The Burland Lithographic Co.).
      1 map : col. ; 49 x 95 cm.
      Accompanies: Report on Hudson's Bay and some of the lakes and rivers to the west of it, 1879-80 / by Robert Bell. Montréal, Quebec : Dawson Brothers, 1881. p. 38C-56C. This report is part C of Report of progress for 1879-80 / Geological and Natural History Survey of Canada.
      Includes index to trees.

(4G4; 7B11)

33.
   [Map showing southern bank of the St. Lawrence River between St. Nicolas and Rivière Ouelle]. -- Scale [ca. 1:63 360]. -- [ca. 1860].
      1 map on 2 sheets : ms., col. ; 47 x 229 cm, on sheets 49 x 119 cm.
      Includes notes in pencil mainly concerning railway line.
      Ms. addition: "Scale 1 inch to 1 mile".
      Map restored. Original folded dimensions, 21 x 15 cm.

(1B7; 5B2a; 5D1f)

34.
   Les Monts Auriol et leurs environs, Yukon / Ministère des mines et des relevés techniques. -- Scale 1:250 000. "4 milles (approximativement) = 1 pounce" (W 138°00' -- W 137°00/N 61°00'-- N 60°30'). -- [Ottawa, Ont.] : Ministère des mines et des relevés techniques, 1951.
      1 map : col. ; 23 x 23 cm, on sheet 50 x 35 cm.
      "La présente carte montrant les Monts Auriol, ainsi désignés en l'honneur de M. Vincent Auriol, Président de la République Francaise, est présentée au Président en vue de commémorer sa visite au Canada en avril 1951."
      Inset: Les Monts Auriol. 6 x 6 cm.

(3B4; 4F7; 5D1h; 7B3)

35a.
   Newfoundland and Labrador official road map. -- Scale [ca. 1:950 400] and [ca. 1:3 168 000] (W 68°00'--W 52°45'/N 61°00'--N 46°30'). -- St. John's. Nfld. : Dept. of Tourism, [1976?].
      2 maps : both sides, col. ; 59 x 58 cm, and 27 x 29 cm, on sheet 62 x 61 cm, folded to 20 x 10 cm.
      Panel title.
      "Published by Newfoundland Department of Tourism ... by authority of ... Department of Transportation and Communications."
      Index of cities, towns and settlements, distance chart. List of provincial parks and beaches with facilities.
      On recto: Province of Newfoundland. Scale [ca. 1:950 400]. 59 x 58 cm. Insets: Transportation routes, Province of Newfoundland. Scale [ca. 1:10 137 600] — St. John's — Corner Brook — Grand Falls — Gander.
      On verso: Labrador. Scale [ca. 1:3 168 000]. 27 x 29 cm — Avalon and Bonavista peninsulas. Scale [ca. 1:506 880]. 50 x 29 cm — Ill.

(3B4; 4F7; 5D1b; 7B3)

35b.

Province of Newfoundland. -- Scale
[ca. 1:950 400] (W 60°00'--W 52°45'/N 52°00'--
N 46°30'). -- St. John's, Nfld. : Dept. of
Tourism, [1976?].
1 map : col. ; 58 x 59 cm, on sheet 62 x 81 cm,
folded to 20 x 10 cm.
Panel title: Newfoundland and Labrador
official road map.
"Published by Newfoundland Department of
Tourism ... by authority of ... Department of
Transportation and Communications."
Index of cities, towns and settlements,
distance chart. List of provincial parks and
beaches with facilities.
Insets: Transportation routes, Province of
Newfoundland. Scale [ca. 1:10 137 600] -- St.
John's — Corner Brook — Grand Falls — Gander.
On verso: Labrador. Scale [ca. 1:3 168 000].
27 x 29 cm — Avalon and Bonavista peninsulas.
Scale [ca. 1:506 880]. 50 x 29 cm — Ill.

*(Example 35 illustrates the general application
at the beginning of area 1, 3 and 5.)*

36.

Niagara-on-the-Lake. -- Scale [1:24 000].
1 in. = 2000 ft. -- St. Catharines [Ont.] :
Proctor & Redfern Limited, 1970.
1 map : col. ; 66 x 59 cm.
Map shows the area municipality of
Niagara-on-the-Lake.
At head of title: The Regional Municipality
of Niagara, area municipality.
"Source of information: Base map derived
from the Surveys and Mapping Branch, Department
of Energy, Mines and Resources."
Includes legend.

(7B6)

37.

Nürnberg, Fürth / Falk-Verlag. -- 22. Aufl. --
Scale 1:16 000-1:26 000 ; hyperbolic proj. --
Hamburg [Germany] : Falk-Verlag, 1969.
1 map : col. ; 56 x 79 cm, folded to 21 x 11 cm.
-- (Falk-Plan ; no. 122)
In German; legend in German, English and French.
Title on spine of cover: Falk-Plan von
Nürnberg.
Accompanied by location plan and street index
attached to cover: [Nürnberg, Fürth]. Scale
[ca. 1:100 000]. 17 x 20 cm.
Insets: Innenstadt Nürnberg. Scale 1:11 000.
17 x 13 cm — Burgfarrnback (Fürth). Scale
1:30 000. 5 x 9 cm — [Kaftshof]. 9 x 10 cm —
[Reichelsdorf-Muhlhof]. 10 x 13 cm.

(Note: *Publisher lists offices in Hamburg,
Berlin, Den Haag, and London.*)

(3B3; 3C1; 4C5; 7B2; 7B4)

38.

Outaouais québecois -- Scale [ca. 1:300 000]
(W 77°55'--W 74°15'/N 47°15'--N 45°20'). --
[Hull, Quebec] : Société d'aménagement de
l'Outaouais, 1976.
1 map : col. ; 73 x 85 cm, folded to 25 x 11 cm.
Panel title: Outaouais Québec.
Inset: Divisions administratives de la zone
métropolitaine de Hull. 16 x 20 cm.
On verso: Hull-Ottawa et environs. Scale
[ca. 1:50 000]. 41 x 47 cm — L'Outaouais
québécois dans le nord-est américain. Scale
[ca. 1:5 500 000]. 23 x 21 cm — Centres
touristiques de la Société d'aménagement de
l'Outaouais. 23 x 21 cm.
Regional and tourist indexes, places of
interest, text and ill.

(1B1a; 5D1h)

39.

Pictorial map of the world. -- Scale
[ca. 1:82 500 000] (W 180°--E 180°/N 80°--S 60°).
-- Bournemouth, England : G.J. Hayter & Co.,
[196-?].
1 jigsaw (125 pieces) : col., plywood ;
25 x 40 cm, in box 27 x 42 x 3 cm. -- (Victory
geographic puzzles ; 7486)
Title from container.
Title block reads: "Victory plywood jig-saw
puzzle pictorial maps of the world, 7486."
Intended audience: For age 7 upwards.

(0.4; 7B14; AACR 2 10.5)

40.

Plan of the township of Hart, Algoma District,
1885 / surveyed by B.J. [i.e. B.G.] Saunders. --
Scale [1:31 680]. 40 chains to 1 in. --
[Toronto, Ont. : Dept. of Crown Lands], 1885.
1 map ; 44 x 36 cm.
Accompanies: Report of the Commissioner of
Crown Lands of the Province of Ontario for the
year 1885. Appendix no. 28, p. 41.

(0F1)

41.

Plan of the village of Cardinal, showing
proposed system of sewerage / designed by
Willis Chipman from surveys and levels by A.J.
McPherson. -- Scale [1:2 400]. 200 ft. to 1 in.
-- 1900.
1 map : ms., col. ; 85 x 80 cm.
Bottom right corner: "Brockville, September
27th, 1900".
Ink drawing.

(1F4; 1F12; 7B9)

42.

Plan of Yellowknife Settlement, Northwest
Territories / compiled from official surveys by
R.W. Clark and C.A.R. Lawrence. -- 2nd ed. --
Scale [1:1 920]. 160 ft. to 1 in. -- Ottawa

[Ont.] : Surveyor General's Office, 1959.
  1 map : photocopy ; 83 x 57 cm.
  Revisions to 1959.
  "Approx. lat. 72°28' ; approx. long.
114°20.6'."
  Diazo print.
  "Plan no. 40273 Legal Surveys and
Aeronautical Charts Division Dept. of Mines and
Technical Surveys, Ottawa."

(2B1; 5C1 Application; 7B8)

43.
  Plan shewing position of beach lot on the
River Ottawa applied for by Perley and Pattee /
W.R. Thistle. -- Scale [1:792]. 1 chain to 1
in. -- 1869.
  1 plan : ms., col., tracing linen ; 48 x 40
cm.
  Provenance: Canada, Department of Public
Works.
  "Note - The figures in red indicate the depth
of water 20th July 1869. On the same date of
1868 the water was 4 1/3 feet lower."

(5C4)

44.
  Province of Newfoundland / Canada, Department
of Mines and Resources, Mines, Forests and
Scientific Services Branch, Surveys and Mapping
Bureau. -- Scale 1:633 600 (W 59°30'--W 52°30'/
N 52°00'--N 46°30'). -- [Ottawa, Ont. : Surveys
and Mapping Bureau], 1949.
  1 map : 97 x 82 cm.
  Shows Dominion electoral district boundaries,
1949.
  Inset: Coast of Labrador. Scale
  1:4 752 000 (W 68°--W 50°/N 60°--N 50°).
28 x 24 cm.

(1F3; 7B1a)

45.
  Quebec, avec plans (centre-ville) de Montréal
et de Québec = Quebec, featuring city area maps
of Montréal and Quebec City. -- Scale
[1:1 100 000] and [ca. 1:133 000] (W 79°30'--
W 64°10'/N 49°15'--N 44°50'). -- Toronto, Ont. :
Creative Map Sales Ltd., c1976.
  1 map : both sides, col. ; 41 x 84 cm, folded
to 22 x 10 cm.
  Panel title.
  Copyright, Creative Sales Corp., 1972-1976.
  Map continues on verso at scale
[ca. 1:133 000].
  Insets: Région de Chibougamau = Chibougamau
area. Scale [1:5 000 000]. 7 x 10 cm —
Northern New Hampshire & Vermont. 9 x 10 cm —
Quebec downtown. 7 x 22 cm — Centre de
Montréal = Central Montréal. 7 x 15 cm —
Province of Quebec. 17 x 28 cm.
  Includes text and ill., time and distance
chart, and partial list of location of cities
and towns.

(1B8a; 1B8b Application 1; 1D1; 7B3)

46.
  Quebec main electric transmission systems and
principal power generating developments /
Department of Energy, Mines and Resources,
Energy Development Sector = Québec principaux
réseaux de transport d'énergie électrique et
principaux aménagements électrogènes / Ministère
de l'énergie, des mines et des ressources,
Secteur de l'exploitation de l'enérgie. -- Scale
[ca. 1:2 000 000] (W 84°--W 57°/N 62°--N 45°).
-- Ottawa [Ont.] : [Energy Development Sector],
1975.
  2 maps : back to back, col. ; 99 x 86 cm each.
  English and French versions of same map.
  Base map produced and printed by the Surveys
and Mapping Branch, Dept. of Energy, Mines and
Resources, Ottawa.
  Insets, scale [ca. 1:506 880]: Québec.
16 x 15 cm — Isle Maligne-Chicoutimi. 16 x 18
cm — Trois Rivières-Shawinigan. 16 x 18 cm —
Montréal. 16 x 18 cm.
  Includes legend.

(1F11)

47.
  Régions électorales : districts électoraux
projetés : [Québec] / rédigée par la Division
de la cartographie, Service de la géographie,
Ministère des transports du Québec en
collaboration avec la Commission permanente de
la réforme des districts électoraux. -- Scale
1:2 500 000 ; conformal conical proj. with two
standard parallels 46° and 60° (W 84°--W 52°/
N 63°--N 45°). -- [Québec, Québec] : Commission
permanente de la réforme des districts
électoraux, 1977.
  1 map : col. ; 85 x 70 cm, in box 25 x 20 x 8
cm. -- (Cinquième rapport et annexe
cartographique / Commission permanente de la
réforme des districts électoraux ; Régions
électorales 1)
  Inset: Iles de la Madeleine. Scale 1:570 000.
20 x 20 cm.
  Includes a legend.
  Accompanies report: Cinquième rapport de la
Commission permanente de la réforme des
districts électoraux. Québec [Québec] :
Commission permanente de la réforme des
districts électoraux, 1978. 167 p.

(1E6; 5D5; 6E1)

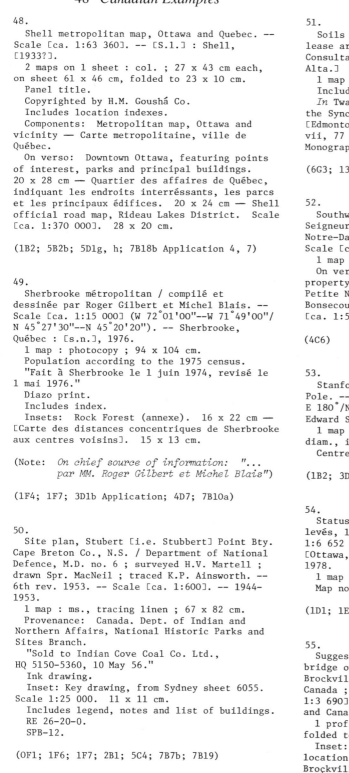

48.
    Shell metropolitan map, Ottawa and Quebec. --
Scale [ca. 1:63 360]. -- [S.1.] : Shell,
[1933?].
    2 maps on 1 sheet : col. ; 27 x 43 cm each,
on sheet 61 x 46 cm, folded to 23 x 10 cm.
    Panel title.
    Copyrighted by H.M. Goushá Co.
    Includes location indexes.
    Components: Metropolitan map, Ottawa and
vicinity — Carte metropolitaine, ville de
Québec.
    On verso: Downtown Ottawa, featuring points
of interest, parks and principal buildings.
20 x 28 cm — Quartier des affaires de Québec,
indiquant les endroits interréssants, les parcs
et les principaux édifices. 20 x 24 cm — Shell
official road map, Rideau Lakes District. Scale
[ca. 1:370 000]. 28 x 20 cm.

(1B2; 5B2b; 5D1g, h; 7B18b Application 4, 7)

49.
    Sherbrooke métropolitan / compilé et
dessinée par Roger Gilbert et Michel Blais. --
Scale [ca. 1:15 000] (W 72°01'00"--W 71°49'00"/
N 45°27'30"--N 45°20'20"). -- Sherbrooke,
Québec : [s.n.], 1976.
    1 map : photocopy ; 94 x 104 cm.
    Population according to the 1975 census.
    "Fait à Sherbrooke le 1 juin 1974, revisé le
1 mai 1976."
    Diazo print.
    Includes index.
    Insets: Rock Forest (annexe). 16 x 22 cm —
[Carte des distances concentriques de Sherbrooke
aux centres voisins]. 15 x 13 cm.

(Note: *On chief source of information: "...*
        par MM. Roger Gilbert et Michel Blais")

(1F4; 1F7; 3D1b Application; 4D7; 7B10a)

50.
    Site plan, Stubert [i.e. Stubbert] Point Bty.
Cape Breton Co., N.S. / Department of National
Defence, M.D. no. 6 ; surveyed H.V. Martell ;
drawn Spr. MacNeil ; traced K.P. Ainsworth. --
6th rev. 1953. -- Scale [ca. 1:600]. -- 1944-
1953.
    1 map : ms., tracing linen ; 67 x 82 cm.
    Provenance: Canada. Dept. of Indian and
Northern Affairs, National Historic Parks and
Sites Branch.
    "Sold to Indian Cove Coal Co. Ltd.,
HQ 5150-5360, 10 May 56."
    Ink drawing.
    Inset: Key drawing, from Sydney sheet 6055.
Scale 1:25 000. 11 x 11 cm.
    Includes legend, notes and list of buildings.
    RE 26-20-0.
    SPB-12.

(OF1; 1F6; 1F7; 2B1; 5C4; 7B7b; 7B19)

51.
    Soils map of a portion of the Syncrude no. 17
lease area, Alberta / mapped by Pedology
Consultants. -- Scale 1:24 000. -- [Edmonton,
Alta.] : Syncrude Canada Ltd., 1978.
    1 map : col. ; 56 x 68 cm.
    Includes legend.
    In Twardy, A.G. Soil survey of a portion of
the Syncrude lease 17 area, Alberta.
[Edmonton, Alta.] : Syncrude Canada Ltd., 1978.
vii, 77 p. ; 28 cm. (Environmental Research
Monograph ; 1978-1).

(6G3; 13E)

52.
    Southwesterly portion of Lucerne-in-Quebec :
Seigneurie de la Petite Nation de la Paroisse de
Notre-Dame-de-Bonsecours, Dominion of Canada. --
Scale [ca. 1:6 857]. -- [S.1. : s.n.], 1931.
    1 map ; 56 x 43 cm.
    On verso: General map of Seigniory club
property, Lucerne-in-Quebec, Seigneurie de la
Petite Nation de la Paroisse de Notre-Dame-de-
Bonsecours, Dominion of Canada. Scale
[ca. 1:59 633]. 56 x 43 cm.

(4C6)

53.
    Stanford's map of countries round the North
Pole. -- Scale [ca. 1:13 260 000] (W 180°--
E 180°/N 90°--N 50°). -- London [England] :
Edward Stanford, 1876.
    1 map : col., mounted on linen ; 62 cm in
diam., in folder 18 x 13 cm.
    Centred on the North Pole.

(1B2; 3D1b Application 2; 5C5; 5D1c; 7B8)

54.
    Status of surveys, 1978 : [Canada] = Type de
levés, 1978 : [Canada]. -- Scale [ca.
1:6 652 800] (W 168°--W 52°/N 88°--N 40°). --
[Ottawa, Ont.] : Canadian Hydrographic Service,
1978.
    1 map : col. ; 80 x 105 cm.
    Map no.: M-270.

(1D1; 1E6; 7B19)

55.
    Suggested highway and single track railway
bridge over the St. Lawrence River near
Brockville / Department of Railways and Canals,
Canada ; W. Chase Thomson. -- Scale [ca.
1:3 690]. -- [Ottawa, Ont. : Dept. of Railways
and Canals], 1927.
    1 profile : photocopy ; on sheet 30 x 78 cm,
folded to 33 x 20 cm.
    Inset: Plan showing suggested bridge
location over St. Lawrence River, near
Brockville. Scale [ca. 1:92 160]. 14 x 17 cm.
    Shows three sections of deck spans, arch span
and trestle.

In Appendix "B" of Report on terminal for use with Welland ship canal. Plate no. 17.

(1F3; 1F6; 5B1a Application; 13E)

56.
    Territoires de ventes de firmes privées : carte synthese no. 1 [Québec] -- Scale [ca. 1:1 013 760] (W 80°--W 65°/N 52°--N 45°). -- [Québec, Québec] : Bureau de recherches économiques, 1965.
    1 map : photocopy ; 82 x 114 cm.
    Working map.
    Provenance: Eric William Kierans, 1914- [Papers].
    Diazo print.
    Insets: Montreal. Scale [ca. 1:506 880]. 8 x 11 cm — [Ile d'Anticosti]. Scale [ca. 1:2 027 520]. 9 x 12 cm — [Iles de la Madeleine]. 9 x 7 cm — Quebec. Scale [ca. 1:506 880]. 8 x 6 cm.

(1E1)

57.
    Town of Preston, with views of principal business buildings. -- Scale indeterminable. -- [S.l : s.n., 1979] (Ottawa, Ont. : National Map Collection, Public Archives of Canada).
    1 microfiche : negative ; 10 x 14 cm.
    Original bird's-eye view published ca. 1900 by Howell Lith. Co., Hamilton, Ont.
    Low reduction.
    National Map Collection microfilm no. 12358.

(0.4; 3B1e; 5B30; 7B10b; AACR 2 11.5D3)

58.
    Tunison's railway, distance and township map of Ontario. -- Scale [ca. 1:829 440] (W 84°00'--W 75°00'/N 46°30'--N 42°00'). -- London, England ; Toronto [Ont.] : H.C. Tunison, 1897.
    1 map : col. ; 66 x 87 cm.
    Insets: Map of north-westerly part of Ontario. Scale [ca. 1:2 644 590]. 21 x 42 cm — London and environs. 7 x 9 cm — Hamilton and environs. 7 x 10 cm — Eastern part of Ontario. 11 x 16 cm — Ottawa and environs. 8 cm diam. — Toronto and environs. 8 x 12 cm.

(Note: *Places listed are: London, England; Toronto, Canada; Sydney, Australia.*)

(4C5)

59.
    United States of America national parks and monuments, national forests; Indian reservations, wildlife refuges, public lands and historical boundaries / prepared and published at the direction of the Congress of the United States ; compiled by the Geological Survey in cooperation with the Bureau of Land Management, United States Department of the Interior. -- Scale 1:3 168 000 ; Albers equal-area proj. based on parallels 29 1/2° and 45 1/2° (W 126°--W 64°/N 49°--N 23°). -- Washington, D.C. : U.S. Dept. of the Interior, 1964 (1965 printing).
    1 map : col. ; 95 x 153 cm.
    Insets: Alaska. Scale 1:12 672 000 (W 176°--W 116°/N 70°--N 48°). 18 x 32 cm — Principal islands of Hawaii. Scale 1:6 366 000 (W 160°20'--W 154°00'/N 22°20'--N 18°40'). 7 x 11 cm — Canal Zone. Scale 1:3 168 000 (W 81°--W 78°/N 10°--N 8°). 7 x 11 cm — Puerto Rico and Virgin Islands. Scale 1:3 168 000 (W 68°00'--W 64°20'/N 19°00'--N 17°30'). 6 x 13 cm — [Key map]. (W 120°--W 60°/N 80°--S 20°).

(3C2; 4G4)

60.
    Ville de Chicoutimi, plan général / conception graphique Angers et Laforge. -- Scale [ca. 1:20 000] (W 71°07'30"--W 71°00'45"/N 48°27'30"--N 48°23'00"). -- Chicoutimi [Québec] : Services techniques, 1977.
    1 map : col. ; 41 x 44 cm, folded to 22 x 10 cm.
    Panel title: Chicoutimi à la carte.
    On verso, scale [ca. 1:1 013 760], 8 x 12 cm each: Autour de la Métropole = Around the Metropolis — De Chicoutimi vers le nord = Up north from Chicoutimi — Valsez avec "Marjolaine" = A cruise on the beautiful Saguenay — Sur les rives du Saguenay = On the shores of the Saguenay — Deux jours au lac St-Jean = A two day tour around the lake St-Jean region — Saguenay-Lac St-Jean-Chibougamau. Scale [ca. 1:500 000]. 18 x 42 cm.
    Index to streets, main buildings and parks.

(1F12; 7B4; 7B18b Application 7)

61.
    Winnipeg 1884. -- Scale indeterminable (W 97°10'00"--W 97°06'30"/N 49°55'00"--N 49°52'40"). -- [Winnipeg, Man. : Hovmand Galleries, 1975].
    1 view : col. ; 48 x 70 cm.
    "Entered according to Act of Parliament of Canada in the year 1884 by W.C. Fonesca in the office of the Minister of Agriculture."
    Facsimile map.
    Insets: 21 insets of local buildings and one perspective view of Winnipeg, 1871.

(3B1e; 5B1a Application; 7B9)

62.

Winnipeg, Port Arthur, Fort William, Brandon.
-- Scales differ. -- Canada : B.A. [Oil
Company], c1962
    3 maps on 1 sheet : both sides, col. ; on
sheet 61 x 81 cm, folded to 21 x 11 cm.
    Panel title.
    "Copyright by Rolph Clark Stone Ltd., Toronto,
1962."
    On recto: Metropolitan Winnipeg. Scale [ca.
1:36 200]. 47 x 49 cm.
    On verso: Port Arthur, Fort William. Scale
[ca. 1:30 000]. 58 x 48 cm — Brandon. Scale
[ca. 1:18 100]. 30 x 27 cm.
    Includes street indexes, tourist information.

(4C1; 4F6)

63.

Yukon Territory / Department of Energy, Mines
and Resources, Surveys and Mapping Branch. --
Scale 1:1 000 000 ; Lambert conformal conic
proj. with standard parallels 61°40' and 68°20'
(W 144°--W 121°/N 71°--N 60°). -- Ottawa
[Ont.] : [Surveys and Mapping Branch], 1972.
    1 map : col. ; 120 x 89 cm.
    Inset: Magnetic diagram, 1970. 19 x 14 cm.
    Map no.: MCR 25.

(1F3; 3C2; 7B19)

64.

    94 G: Trutch, British Columbia / prepared
by the Army Map Service (AM), Corps of
Engineers. -- Ed. 1, AMS. -- Scale 1:250 000.
Vertical scale 1:125 000. Vertical
exaggeration 2:1 ; universal transverse
Mercator proj. (W 124°--W 122°/N 58°--N 57°). --
Washington, D.C. : Army Map Service, Corps of
Engineers, 1956.
    1 model : col., plastic ; 45 x 49 cm. --
(Plastic relief maps of western Canada,
1:250 000 : series A502P ; sheet 94 G)
    Universal transverse Mercator grid.
    "Constructed and molded in 1956 from Canada,
1:250,000 Series A502, Edition 1, Sheet 94 G,
Department of Mines and Technical Surveys,
1952."
    Includes legend.
    Item to be used for display by National Map
Collection.

(3B8; 5B1a Application; 5C4; 6G1)

65.

Outdoor recreation, eastern Ontario-western
Quebec = Récréation de plein-air, est de
l'Ontario-ouest du Québec ; Ottawa-Carleton : a
proposal for outdoor recreation = Ottawa-
Carleton : un projet de récréation de plein-air.
-- Scale 1:100 000 and 1:250 000. -- [Ottawa,
Ont.] : Regional Municipality of Ottawa-
Carleton, Planning Department, 1977.
    2 maps : both sides, col. ; 79 x 94 cm each.
-- (River corridors and conservation areas

project ; exhibit 3A = Projet d'étude, couloirs
fluviaux et zones de conservation ; document 3A)
(River corridors and conservation areas project
; exhibit 3B = Projet d'étude, couloirs fluviaux
et zones de conservation ; document 3B)
    Accompanies report: Ottawa and Rideau rivers
and shores in Ottawa-Carleton. [Ottawa, Ont.] :
Planning Dept., Regional Municipality of Ottawa-
Carleton, 1978. xi, 211 p. ; 28 x 44 cm.
    Includes legend.

(1G3)

66.

Township of Gloucester : schedule A, sheet
one to zoning by-law no. 26 of 1960 / McRostie &
Associates Ltd. -- As rev. by by-law no. 19, of
1963. -- Scale [ca. 1:35 000]. -- [S.l. : s.n.,
1979] (Ottawa, Ont. : National Map Collection,
Public Archives of Canada)
    1 microfiche : negative ; 10 x 14 cm.
    "Base map of township by National Capital
Commission."
    Low reduction.
    National Map Collection microfilm no.: 13665.

(2B1; 4G1)

67.

Metropolitan Vancouver, British Columbia,
Canada / produced by Webb & Knapp (B.C.) Ltd.,
with the co-operation of the cities and
municipalities forming the Greater Vancouver
metropolitan area. -- Scale [ca. 1:63 360]. --
Vancouver, B.C. : Webb & Knapp, 1958.
    1 remote sensing image : col. ; 48 x 63 cm.
    Photomap with symbols overprinted in various
colours.
    Photographic inset showing view of
metropolitan Vancouver.
    Includes keys to different areas.

(5B1a Application; 7B10a)

68.

Map, London Ontario, Canada / compiled
exclusively by Pathfinder Air Surveys Ltd. --
Scale [1:16 200]. 350 ft. to 1 in. (W 81°21'05"
-- W 81°09'35"/N 43°01'25"--N 42°56'15"). --
Ottawa [Ont.] : Pathfinder ; London [Ont.] :
Western News Co. [distributor], c1963.
    1 map : col. ; 59 x 93 cm, folded to 21 x 13
cm.
    Produced from aerial photographs.
    Includes legend and index of street names,
churches, schools, cemeteries, parks, hospitals,
miscellaneous landmarks and industries.
    Inset: Key map of London, the forest city,
Ontario. Scale [ca. 1:75 600]. 16 x 20 cm.

(4E1)

69.
   Canada showing the Bay stores. -- Scale
1:6 336 000 (W 145°--W 50°/N 81°--N 40°). --
[Winnipeg? Man.] : Hudson's Bay Co., 1979
printing.
   1 map : col. ; 65 x 92 cm.
   "Base map produced and printed by the Surveys
and Mapping Branch, 1970."
   Includes legend.

(4F6)

70.
   Map of region Niagara / prepared by Fred P.
Paul. -- Rev. 1977 / J.B. Phillips. -- Scale
[ca. 1:145 000] (W 79°35'58"--W 78°53'55"/
N 43°16'40"--N 42°50'03"). -- St. Catharines,
Ont. : Region Niagara Tourist Council, 1977.
   1 map ; 37 x 54 cm.
   Title at lower left: Regional Municipality of
Niagara.
   Includes legend and tourist information.

(2C1)

71.
   Canada / Department of Mines and Technical
Surveys, Surveys and Mapping Branch. -- Scale
1:8 870 400. 140 miles to 1 in. (W 141°--W 50°/
N 83°--N 41°). -- Ottawa [Ont.] : Department of
Citizenship and Immigration, 1951 (1954
printing).
   1 map : col. ; 52 x 67 cm, folded to
28 x 11 cm.
   Text in English, French, Dutch, German,
Spanish and Polish on verso.
   Includes legend.

(4G4)

72.
   Reader's Digest great world globe. -- Scale
1:41 849 600. -- Chicago, Ill. : Replogle Globes,
Inc., [ca. 1967].
   1 globe : col., mounted on metal stand ; 31 cm
in diam.
   Raised relief.

(5B1a Application; 5C5; 5D4)

Australian examples

200.
   Australia, geographic map / produced by the
Division of National Mapping, Dept. of Minerals
and Energy. -- 3rd ed. -- Scale 1:2 500 000. --
Canberra, A.C.T. : D.N.M., 1975.
   1 map on 4 sheets : col. ; 160 x 189 cm.
   Relief shown in hypsometric tints, contours,
hill shading and spot heights. Ocean depths
shown by bathymetric contours and tints.
   Marginal title.
   Panel title: Australia.

(1B8b Application 1; 5B2b; 7B1a Application;
7B3; 7B4)

201.
   Australia 1:500 000, Sydney, metallogenic map
/ issued under authority of the Minister for
Dept. of Mineral Resources and Development. --
1st ed. -- Scale 1:500 000. -- Sydney [N.S.W.] :
Geological Survey of N.S.W., Dept. of Mineral
Resources and Development, 1980.
   1 map : col. ; 66 x 55 cm., folded to
17 x 15 cm.
   Map extends from Nowra to Cape Hawke.
   Accompanied by: Sydney 1:500 000 metallogenic
map : mine data sheets / by G.R. McIlveen.
Sydney [N.S.W.] : Geological Survey of New South
Wales, 1980. 168 p. ; 24 cm.
   Ancillary map: Structural units.

(1B8b Application 2; 5D1h; 5E1c; 7B1a; 7B11;
7B18b Application 5)

202.
   The wine producing area of Australia /
Australian Wine Board. -- Scale indeterminable.
-- Adelaide [S. Aust.] : Australian Wine Board,
1977.
   1 map : col. ; 67 x 91 cm.
   Insets: Western Australia — Swan Valley —
Barossa Valley — Coonawarra-Padthaway —
Clare-Watervale — Mudgee — Hunter Valley —
Forbes Cowra-Griffith — Leeton — Drumborg —
Mildura-Robinvale — Swan Hill — Glenrowan-
Milawa — Corowa-Rutherglen — Great Western-
Avoca — Goulburn Valley — Lilydale —
Langhorne Creek — Southern districts —
Adelaide metropolitan — Riverland.
   Key map showing location of inset maps.

(3B1e; 5B1a; 7B18b Application 5)

203.
   Map of Sepik District / compiled by the
Department of Lands and Mines. -- Scale
1:500 000. -- Rabaul [New Guinea] : Dept. of
Lands & Mines, 1935.
   1 map : photocopy ; 78 x 87 cm.
   Shows Sepik district of New Guinea annotated
with route of an oil exploration party showing
daily stages of journey and recording locations
of attacks on personnel.
   New boundary of uncontrolled area (Gazette no.
546) marked in red ink.
   Blueprint.

(3B1e; 5B1a Application; 7B18b Application 5)

204.
Tour planning map showing road conditions for Victoria, N.S.W. and Queensland ; Interstate routes between Adelaide, Melbourne, Sydney, Brisbane and Cairns ; Principal roads in Australia / prepared and published by the Royal Automobile Club of Victoria (RACV) Ltd. -- 21st ed. -- Scales differ. -- Melbourne [Vic.] : RACV, [1977].
    3 maps on 1 sheet : col. ; sheet 46 x 70 cm.
    Panel title.
    On recto: Bitumen and gravel roads Adelaide, Melbourne, Sydney.
    On verso: Sydney, Brisbane, Cairns. Scales differ — Principal roads in Australia. Scale ca. [1:15 000 000].

(1B8b Application 1; 3B5; 5B2b; 7B3; 7B18b Application 3)

205.
A sketch of Harbour Grace and Carbonere in Newfoundland / by Jas. Cook. -- Scale [ca. 1:28 950]. -- 1762.
    1 map : ms., col., mounted on linen ; 38.0 x 30.5 cm., on sheet 42.0 x 36.0 cm., folded to 23.0 x 36.0 cm.
    Nautical chart of Harbour Grace area, east coast of Bay de Verde Peninsula.
    Housed in Manuscript Collection.
    In manuscript notebook entitled: Description of the sea coast of Nova Scotia. The volume is bound in morocco and bears the cover title: Captain Cook's exploration of Newfoundland. Original notebook with maps. 1762.

(1B1a; 1F1; 4F9; 5D1a; 13E1)

206. No example

207.
A chart of the islands of Jersey and Guernsey, Sark, Herm and Alderney, with the adjacent coast of France / by L.S. De La Rochette. -- Scale [ca. 1:250 000]. Nautic leagues 20 to a degree. -- London : Engraved & published ... by W. Faden ... Charing Cross, Jany. 1st 1781.
    1 map : col. ; 68 x 50 cm.
    Nautical chart. Shows also redoubts and guardhouses along the French coast. Notes on tidal movements.
    Title from spine.
    In General atlas of modern geography. -- Vol. 1. Map no. 22.

(3B2 Application; 4F10; 7B3; 13E1)

208.
[Sketch of St. John River, New Brunswick / by James Cook]. -- Scale [ca. 1:170 380]. -- [1762].
    1 map : ms., col., mounted on linen ; 71.0 x 22.2 cm., on sheet 74.0 x 24.0 cm., folded to 36.5 x 24.0 cm.

Shows the mouth of the St. John River to approximately sixty-five miles inland. River soundings, sandbanks, portages and settlements marked.
    Housed in Manuscript Collection.
    In manuscript notebook entitled: Description of the sea coast of Nova Scotia. The volume is bound in morocco and bears the cover title: Captain Cook's exploration of Newfoundland. Original notebook with maps. 1762.

(1B7; 1F17; 3B1d; 4F7; 5C1 Application; 7B20a; 13E1)

209.
[Sketch of St. John River, New Brunswick / by James Cook]. -- Scale [ca. 1:170 380]. -- [1762].
    1 map ; 73 x 22 cm.
    Shows the mouth of the St. John River to approximately sixty-five miles inland.
    Facsim. of manuscript map coloured, and mounted, contained in manuscript notebook entitled: Description of the sea coast of Nova Scotia.
    The original manuscript is bound in morocco and bears the cover title: Captain Cook's exploration of Newfoundland. Original notebook with maps. 1762.
    Original in Manuscript Collection. (ANL MS 5)

(5D1b; 7B7a; 7B20a; 11A)

210.
Melbourne and Metropolitan Board of Works : [Melbourne] scale 1600 feet to 1 inch. -- Scale [1:19 200]. 1600 ft. to 1 in. -- [Melbourne] : M.M.B.W., [196-?]-
        maps ; 25 x 51 cm.
    Street plans of Melbourne and suburbs.
    Contents: 5F: Municipalities of Chelsea, Dandenong and Mordialloc. [196-?].
              5G: Municipalities of Chelsea, Cranbourne and Dandenong. [196-?].
              6F: Municipalities of Dandenong, Moorabbin, Mordialloc and Sandringham. [196-?].
              7F: Municipalities of Brighton, Sandringham, Moorabbin and Oakleigh. [196-?].

(1B3; 1E6; 1F13; 4C6; 4D5; 4F7; 7B18b Application 4)

211.
[Western Australia forestry series 1:63 360] / B.J. Beggs. -- Scale 1:63 360. -- [Perth, W.A.] : Western Australia Forests Dept., 1963-
        maps : col. ; 65 x 96 cm. or smaller.
    Series extends over southwestern Western Australia. Shows occurrence of private and public land, forest blocks, karri, regenerated forest, permanent water supply and picnic areas.

Contents: Augusta. Aug. 1973.
           Busselton. July 1971.
           Chudalup. June 1977.
           Collie. Sept. 1968.

(1B7 Application 2; 5D1k)

212.

Fire control map, region 4 - Victoria /
prepared by Golden Fleece Petroleum (for the
Committee of Region 4, Victorian Country Fire
Authority from information supplied by Region 4
Fire Brigade Groups). -- Scale 1:126 720. 2
miles to 1 in. -- [Melbourne, Vic.] : Golden
Fleece Petroleum, 1974.
     1 map : col. ; 119 x 81 cm.
Map shows area from South Australian border
eastwards to Hamilton and from the coast inland
to Rocklands Reservoir. Roads classified by
surface, broad categories of vegetation cover
indicated, fire amenities and water filling
points shown.
     Panel title.
     Title on recto: Fire control map, region 4.
     On verso: Numerical index — Alphabetical
index — Localities and communications —
Aircraft landing grounds.

(1F1; 3B2; 7B4; 7B18b Application 7)

213.

Commonwealth of Australia Petroleum
(Submerged Lands) Act 1967, proclaimed 1st
April 1968 : State of Western Australia
Petroleum (Submerged Lands) Act, proclaimed 1st
April 1968, Section 17 : Identification of
graticular sections in the Adjacent Area /
prepared by the Survey and Mapping Branch, Mines
Dept., Western Australia. -- Scale 1:1 000 000.
-- Perth, W.A. : Dept. of Mines, 1967-
     maps : photocopies ; 64 x 69 cm.
Shows five-minute blocks available for
prospecting in the adjacent offshore waters of
Western Australia. The boundary of the
Adjacent Area, including the boundary of the
internal waters and the seabed boundary agreed
9 Oct. 1972 with Indonesia are marked.
     Diazo prints.
     Contents: SC 51: Timor Island. 30 June
                        1977.
                SC 52: Melville Island. 30 June
                        1977.
                SD 51: Brunswick Bay. Provisional
                        copy. 30 June 1977.
                SD 51: Brunswick Bay. 30 June
                        1977.

(5B28)

214.

Fitzroy Region, Queensland : resources
series / Resources, Information and
Development Branch, Department of National
Development. -- Scales differ. -- Canberra
[A.C.T.] : R.I.D. Branch, Dept. of National
Development, 1965-

maps : col. ; 76 x 56 cm., folded to
19 x 14 cm.
Shows various aspects of the hinterland of
the area between Gladstone and Mackay.

     Climate. -- Scale 1:1 000 000. -- 1965. --
Accompanied by report: Climate / prepared by
the Queensland Regional Office of the
Commonwealth Bureau of Meteorology and E.A.
Fitzpatrick. Canberra [A.C.T.], 1965. 28 p.

     Railways and ports. -- Scale 1:1 000 000. --
1966. -- Accompanied by report: Railways and
ports / prepared by Queensland Dept. of
Railways and Queensland Dept. of Harbours and
Marine. Canberra [A.C.T.], 1966. 14 p.

(4D3; 5B28; 7B11; 13F)

215.

Central coast, NSW, Australia / published by
Gregory's Street Directories, Guides and Maps.
-- 6th ed. -- Scale [ca. 1:25 000]. -- Ultimo,
N.S.W. : Gregory's, 1979.
     1 map : col. ; 198 x 72 cm., on sheet 99 x 72
cm., folded to 25 x 13 cm. -- (Gregory's
district map series ; Map 503)
     Panel title.
     Includes index to localities, places of
interest.
     On verso: Index to streets.

(6G1)

216.

The U.B.D. map of Ipswich / Universal
Business Directories Pty. Ltd. -- Scale
[ca. 1:250 000]. -- South Brisbane : UBD,
[1979].
     1 map : col. ; 62 x 65 cm., on sheet 73 x 99
cm., folded to 29 x 11 cm. -- (Explorer series
map)
     Includes index to streets, advertisements.
     ISBN 0-7261-4787-7.

(4D5; 6B; 8B1)

217.

Port Macquarie & district including maps of
Camden Haven, Wauchope / I. Broadbent Johns
Pty. Ltd. -- Scales differ. -- Wahroonga,
N.S.W. : I. Broadbent Johns, [1979?].
     4 maps on 1 sheet : col. ; 40 x 33 cm., sheet
40 x 58 cm., folded to 20 x 12 cm. -- (R.A.
Broadbent tourist maps)
     Panel title.
     On recto: Port Macquarie. Scale
[ca. 1:25 000]. Inset: Town centre.
     On verso: [Port Macquarie and district].
Scale [ca. 1:400 000] — Camden Haven. Scale
[ca. 1:25 000] — Wauchope. Scale 1:25 000.

(4F7; 7B3)

218.
    Rainfall map of Queensland from telegraphic
reports only / Bureau of Meteorology. -- Scale
1:5 000 000 ; Albers conical equal area proj. --
Brisbane [Qld.] : Bureau of Meteorology, 1976-
        maps : col. ; 54 x 39 cm.
    Contents: June 1980.
                July 1980.
                Aug. 1980.

(3C1)

219.
    The Kingdom of Tonga. -- 2nd ed. -- Scale
[ca. 1:100 000 and ca. 1:2 000 000]. -- Surry
Hills, N.S.W. : Pacific Maps, [1979?].
    8 maps on 1 sheet : col. ; 57 x 75 cm.
    General maps showing principal roads and
settlements of various islands.

    Published in association with the Friendly
Islands Bookshop, Nuku'alofa, with
acknowledgements to the Government of Tonga.
    Accompanied by one map: Nuku'alofa. Scale
[ca. 1:20 000].   18 x 24 cm.   Town plan.
    Extensive descriptive notes, Tongan legends
and ill.
    Components: Tongatapu. Scale [ca. 1:100 000]
-- Tonga. Scale [ca. 1:2 000 000] -- Location of
Tonga -- Niuatoputapu. Scale [ca. 1:100 000] --
Niuafo'ou. Scale [ca. 1:100 000] -- Vava'u Group.
Scale [ca. 1:100 000] -- Ha'apai Group. Scale
[ca. 1:100 000].
    On verso: [Artist's portrayal of Tongan life].
Sketch drawn by Erwin Weber of the Lenati family
of Haveluloto. 1979.

(3B4; 7B9; 7B11; 7B18b Application 4, 7)

British examples

300.
Tirol, Voralberg, Oberbayern, Dolomiten
Touristenkart / Kümmerly & Frey Geographischer
Verlag. -- Scale 1:500,000 (E 9°12'--E 14°00'/
N 48°07'--N 45°24'). -- Bern : Kümmerly & Frey,
[1951?]. -- 1 map : col. ; 62 x 74 cm, folded
to 23 x 13 cm.

Legend in German and Italian.
Title at top left: Touristenkarte von Tirol.
Title at top right: Carta turistica dell'Alto
Adige.

(1B8a; 3D1b Application; 4F7; 7B4)

301.
Lake District : tourist map / Ordnance Survey
of Great Britain. -- [Ed.] C6 [bar, star]. --
Scale 1:63,360. -- Southampton : Ordnance
Survey, [1979], c1966. -- 1 map : col. ;
92 x 78 cm.

Compiled from one inch seventh series sheets,
rev. 1962-64, selected roads rev. 1978.

(2D1; 4F5; 7B7a)

302.
Path of the new planet Astrea for 1846 : the
stars from Bessel's zones, the planets places
by J.R. Hind: as seen in an inverted
telescope. -- 2nd ed. -- (RA 3 hr. 51 min. to
4 hr. 19 min./ Decl. +17°15' to +11°35'). --
London : J. Wyld, 1846. -- 1 map ; 22 x 18 cm.

(3D2; 5B1a Application)

303.
A map of the islands of Scilly : showing all
the rocks and ledges, soundings and bearings,

and the exact places where the Association,
Eagle, Romney and Firebrand were lost / Edmund
Gostelo. -- Scale [ca. 1:63,360]. -- [1707]. --
1 map : ms. ; 42.4 cm in diam., on sheet
43.2 x 56.0 cm.

The ships were lost in 1707.
Ink drawing.
Title in cartouche in the form of an archway,
supported by figures. Scale in baroque
cartouche with figures and faces, surmounted by
a griffin. Arms of the United Kingdom featured.
Outside the circumference of the maps are
drawings of farm labourers at work and domestic,
wild and mythical animals.

(7B18b Application 6)

304.
Survey of the ground fortified on the south
side of the River Tagus / Edward B. Metcalf. --
Scale [1:10,138]. -- 1813. -- 1 map : ms.,
col. ; 53.6 x 96.5 cm.

Area covered is in Estremadura province showing
towns, roads, villages and fortifications.
Received with a letter from Mr. Metcalf to Lt.
Gen. Mann, dated Lisbon 10 Oct. 1813.
Bottom left-hand corner damaged.

(1F1; 7B1a; 7B7b)

305.
G. Bradshaw's map of canals, navigable rivers,
railroads &c., in the midland counties of
England from actual survey : showing the
heights of the ponds on the lines of
navigation from a level of 6ft. 10in. under the
Old Dock Sill at Liverpool from levels taken by
Twyford & Wilson Surveyors & Engineers,

Manchester / engraved and printed by W.R.
Gardner. -- Scale [ca. 1:130,000]. --
Manchester : G. Bradshaw ; London (163 Regent
Street) : to be sold by Jas. Gardner, Feb. 12th
1829. -- 1 map : col., mounted on linen ;
127 x 117 cm, dissected into 48 sheets each
23.6 x 15.6 cm.

Table: Sizes of the largest vessels which the
locks will admit. Dedication: To Thomas
Telford F.R.S.L. & E. President of the
Institution of Civil Engineers this map as a
tribute of respect is by permission dedicated
by his obliged G. Bradshaw.
In slip case with: G. Bradshaw's map of canals
situated in the counties of Lancaster, York,
Derby & Chester ...
Advertised: Price mounted on rollers £2.2s:
in sheets £1.11.6d.

(1B2; 1B14; 1E1; 1E4; 1F1; 3B1d; 4C7; 7B1a; 7B5
Application; 7B21; 8D1)

306.
Plan of Chatham dock and its environs
comprehending Rochester Bridge and Stroode,
Gillingham, Fort Gillingham, Upnor Castle,
Cockham Wood Fort and the adjacent country :
showing lines as they are to be restored this
present year 1779. -- Scale [ca. 1:9,647]. -- 1
map : ms., col. ; 49.1 x 62.8 cm.

Title and profiles in panel to the left of the
plan.
Copied from a survey made in the year 1724.
Bar scale of 4,100 ft. to 5.1 in. Compass
indicator. Two profiles of the line of Chatham,
1756 and 1779.
On verso: Endorsed Chatham Dock and its
environs.
In the same folder is another copy of this plan.

(1B15)

307.
[Ohio River and tributaries from Fort Presque
Isle on Lake Erie to the Potomac River] /
[George Washington]. -- Scale [ca. 1:1,111,334].
-- 1754. -- 1 map : ms. ; 45.6 x 35.4 cm.

Shows place names, forts, settlements, Allegheny
Mountains. A note on the map, enclosed in a
square, describes the French purpose to advance
from Lake Erie via French Creek to the Ohio and
beyond and the British intention to build a fort
at the junction of the Ohio and Monongahela.
Reproduced in: Washington's map of the Ohio /
by Worthington Chauncey Ford. Boston : [s.n.],
1928.
Bar scale of 110 miles to 6.2 in. Magnetic
compass star.
Ink drawing.
On verso: in Gov. Dinwiddie's letter of Jany
29th 1754.

(1E6; 1F17; 7B1a; 7B7c)

308.
Spain and Portugal / ed. by J. Bartholomew. --
Scale 1:2,000,000. -- Edinburgh : Bartholomew,
[1922]. -- 1 map : col. ; 49 x 69 cm, folded to
22 x 13 cm. -- (Bartholomew world survey series)

Bartholomew printed code "A22" in top left
margin dates map from first half of 1922.

(1F1; 4D3; 6B1; 7B9)

309.
Norwich / Geographia. -- New ed. -- Scale
[1:15,840]. 4 in. to 1 mile (E 1°13'--
E 1°24'/N 52°41'--N 52°35'). -- London :
Geographia Ltd., [1980?]. -- 1 map ; 54 x 77
cm, folded to 21 x 14 cm.

Cover title: Norwich street plan with index to
places of interest.
National grid.
Date code: CDLN CAD believed to be 10.79 1.80.
Inset of city centre at [ca. 1:7,920].
ISBN 0 09 201810 6.

(2B1; 7B9; 8B1)

310.
Atlas des nördlichen gestirnten Himmels für den
Anfang des Jahres 1855 / unter Mitwirkung der
Herren E. Schönfeld und A. Krueger, nach der in
den Jahren 1852 bis 1862 auf der königlichen
Universitäts-Sternwarte zu Bonn durchgeführten
Durchmusterung des nördlichen Himmels entworfen
und in Namen der Sternwarte herausgegeben von
dem Director derselben F.W.A. Argelander. --
(Zones +90° to +79°, +80° to +60°, +61° to +40°,
... +20° to +2°; eq. 1855). -- Bonn : Bei Adolph
Marcus, 1863. -- 40 maps ; 63 x 43 cm or
smaller.

Limiting magnitude old scale 10.0.

(1B1a; 3D2; 7B8)

311.
To the Right Honourable the Lords Commissioners
for executing the office of Lord High Admiral
of Great Britain, Ireland &c. this chart of St.
George's Channel &c. surveyed under their
lordships direction by the late Lewis Morris
Esqr. which is now extended by an actual survey
(the whole corrected & much improved) from
Liverpool to Cardiff in the Bristol Channel / is
by permission most respectfully dedicated by
their lordships obedient humble servant William
Morris ; engraved by William Simkins. -- Scale
[ca. 1:380,750]. -- [S.1.] : William Morris, 25
Nov. 1800. -- 1 map ; 85 x 84 cm.

Shows the west coast of England and Wales from
Formby Point to Cardiff; and east coast of
Ireland from Shenez Is. to Carnsore Point.
Based partly on: A chart of the coast of Wales
... between the years 1737 & 1744 ... / by Lewis
Morris, published ... Sept. 30 1748.

"A scale of 10 leagues". Scale approximates 6 miles to 1 in. Latitude and longitude shown in units of 5'.
Dedication to companion volume: Plan of the principal harbours ... a new edition ... / by William Morris. Shrewsbury : Printed for the author by Sandford and Maddocks, 1801. Signed William Morris, Aberstwith [i.e. Aberystwyth] June 1, 1801.
Reference: Marine plans and charts of Wales / O.C. Evans. 1969. No. 106.

(1B4 Application; 4F10; 7B15)

312.
A chart of the coast of Wales in St. George's Channel drawn from an actual survey now lying at the Admiralty Office and made between the years 1737 & 1744 by the order and encouragement of the Lords of the Admiralty / by Lewis Morris ; Nathl. Hill sculp. -- Scale [ca. 1:437,900]. -- [London : Lewis Morris], Sept. 30 1748. -- 1 map ; 50 x 71 cm.

Shows coast of Wales from Orme's Head to Milford.
Dedication: To the Right Honourable the Lords Commissioners for executing the office of Lord High Admiral of Great Britain, Ireland &c. this chart is with the greatest submission presented by their Lordship's most obedient humble servant Lewis Morris.
Latitude shown in units of 10. Longitude not shown.
Oriented with north to right. "A scale of 15 leagues."
"Publish'd by the author according to Act of Parliament."
Note to the reader in companion volume: Plans of harbours ... / by Lewis Morris, 1748.
Signed Lewis Morris, London Feb. 2 1748.
References: Marine plans and charts of Wales / O.C. Evans 1969. No. 105.

Decorative dedication cartouche with vignettes of Welsh industry: "the woollen manufacturing, ship building, husbandry, mineing [sic], timber falling [sic], the fishery."

(4F10; 7B1a; 7B5 Application; 7B8; 7B9; 7B15; 7B20b)

313.
To Sir Watkin Williams Wynn Bart. this map of North Wales is respectfully inscribed / by his obliged and obedient servant John Evans ; engraved by Robert Baugh. -- Scale [ca. 1:204,380]. -- [S.l. : s.n.], March 27th 1797 [i.e. 1802]. -- 1 map ; 61 x 71 cm.
This edition has table of turnpike, cross and intended roads below scale bar.
Additional roads shown.
Scale bar: 20 miles to 6¼ in.
Probably published at Llanymynech by Dr. John Evans in 1802.
Cartographic Journal Vol. 5 no. 2 (1968) p. 140.
View of Conway Castle.

Copies in N.L.W.

1. PA 93 (WB copy) - 1 map : mounted on linen ; 61 x 72 cm folded to 22 x 15 cm.
2. PB 6054 - 1 map : mounted on linen ; 62 x 72 cm folded to 19 x 13 cm.
3. PB 93 (C.M. Elderton copy) - 1 map : hand col., mounted on linen ; 62 x 72 cm folded to 18 x 23 cm.
4. PA 838 - 1 map : mounted on linen ; 61 x 72 cm folded to 22 x 13 cm, in slip case with label: "Sold by W. Faden, Geographer to His Majesty and His Royal Highness the Prince to Wales, No. 5 Charing Cross".

(1B4 Application; 4F2; 7B7a; 7B18b Application 6; 7B20a)

United States examples

400.
   New Zealand land resource inventory worksheet / produced for the National Water & Soil Conservation Organisation by the Water & Soil Division, Ministry of Works & Development. -- Scale 1:63,360. 1 in. to 1 mile (E 166°-- E 179°/S 34°--S 48°). -- [Wellington, N.Z.] : The Organisation, [1975-
      maps : col. ; 53 x 79 cm. or smaller, folded in cover 63 x 49 cm. + legend ([1] leaf, 12 p. ; 61 cm.)
   Geographic coverage complete in 360 maps.
   Relief shown by spot heights and landform drawings.
   "Base map by the Department of Lands and Survey: NZMS 1 series ..."
   Accompanied by errata sheets.
   Includes index to adjoining sheets.

Index to sheets on back cover.
Some maps have errata sheets pasted on.

(4D5; 5E1d; 7B20a)

401.
   Water availability of Randolph County, Alabama / by Gregory C. Lines and Robert V. Chandler ; prepared in cooperation with the United States Geological Survey. -- Scale [ca. 1:126,720] (W 85°40'--W 85°15'/N 33°30'-- N 33°05'). -- University, Ala. : Geological Survey of Alabama, 1975.
   1 map : col. ; 48 x 49 cm., folded in envelope 23 x 31 cm. -- (Map / Geological Survey of Alabama ; 137)
      Shows surface and underground water flow.
      Relief shown by spot heights.

"Base map modified from Alabama Highway
Department maps and field notes."
Accompanied by text: Water availability,
Randolph County, Alabama.
Includes text and location map.

(5D5; 6B1; 6E1; 7B1a; 7B11)

402.
Official city map of Charlotte, Mecklenburg
County, North Carolina / compiled by the
Engineering Division, Department of Public
Works, city of Charlotte ; photography &
mapping by Abrams Aerial Survey Corporation. --
Scale [1:24,000]. 1" = 2,000'. -- [Charlotte] :
The Division, c1974.
    1 map : photocopy ; 104 x 158 cm.
    Alternate title: Charlotte.
    "Grid based on North Carolina rectangular
coordinate system."
    Blue line print.
    Inset: Mecklenburg County, North Carolina.
    "A.A.S.C. 12730."

(5C1 Application; 7B4; 7B8; 7B10a; 7B19)

403.
    New York State map : four sheet, 1:250,000 /
prepared and published by the New York State
Department of Transportation, in cooperation
with the Federal Highway Administration, U.S.
Department of Transportation. -- Rev. in
1979 from Dept. of Transportation 1:24,000
scale quadrangle maps, highway construction
plans, municipal boundary maps, and various
other sources. -- Scale 1:250,000 ;
transverse Mercator proj. (W 79°52'30"--
W 71°45'00"/ N 45°00'00"--N 40°22'30"). --
Albany : N.Y. State Dept. of Transportation :
For sale by the Map Information Unit, N.Y.
State Dept. of Transportation, 1980.
    4 maps : col. ; 95 x 132 cm.
    Shows minor civil divisions.
    "New York transverse Mercator (NYTM) grid."
    Includes sheet index.
    Contents: West sheet — North sheet —
Central sheet — South sheet.

(2B1; 4D4)

404.
    State of Georgia, NASA LANDSAT-1 satellite
image mosaic / prepared and published by the
U.S. Geological Survey in cooperation with
the National Aeronautics and Space
Administration and the Earth and Water
Division of the Georgia Department of
Natural Resources. -- Scale 1:500,000. 1 cm.
equals 5 km. ; Lambert conformal conic proj.
based on standard parallels 33° and 45°
(W 85°--W 81°/N 35°--N 31°). -- Reston, Va. :
The Survey, 1976.
    1 remote sensing image : col. ; 109 x 95 cm.
    Title in lower margin: Georgia satellite
mosaic, 1973-1974.

"Imagery recorded in discrete spectral
bands with multispectral scanner (MSS) on
NASA LANDSAT-1 (formerly ERTS-1). Orbital
altitude 920 km. (570 mi.)."
    "20,000-metre universal transverse
Mercator grid, zones 16 and 17."
    Includes text and 4 diagrams.

(3C2)

405.
    Mare Nectaris and vicinity, lunar plastic
relief map / produced by Army Map Service,
Corps of Engineers. -- Scale 1:5,000,000.
Vertical scale 1:1,000,000. Vertical
exaggeration 5:1. -- Washington : The Service,
[1965]
    1 model : col., plastic ; 31 x 22 x 2 cm.
    "Prepared by the Army Map Service (AM),
Corps of Engineers, U.S. Army, Washington, D.C.,
from Lunar topographic map 1:5,000,000 Army Map
Service, provisional edition, 1961."
    "2-65."

(3B8; 5D3)

406.
    Rand McNally international globe. -- Scale
1:42,000,000 (W 180°--E 180°/N 90°--S 90°). --
[Chicago] : Rand McNally & Co., c1978.
    1 globe : col., plastic mounted on metal
stand ; 31 cm. in diam.
    Raised relief globe. Relief also shown by
shading and spot heights. Depths shown by
shading and soundings.
    Mounted on metal stand by a metal cradle
connected to a circular, movable metal scale
graduated in statute miles and hours of time,
which in turn is connected to a second
circular metal scale graduated in degrees of
latitude.
    Includes plastic indicator to "Set hour over
one city then read time for other cities."

(5B1a Application; 5C5; 5D4; 7B10a)

407.
    Exonymen : [West-Europa] / samenstelling,
P.C.J. van der Krogt ; kartografie, Geogr.
Cart. Inst. Wolters-Noordhoff. -- Scale
1:8,000,000 (W 10°--E 15°/N 62°--N 44°). --
[Groningen? Netherlands : s.n., 1980?]
    8 maps on 1 sheet : col. ; 25 x 18 cm.,
sheet 60 x 86 cm.
    Each map shows major geographic names in the
language of a selected atlas source.
    Includes source citations.
    Contents: 1. Nederlands — 2. Duits —
3. Engels — 4. Frans — 5. Italiaans —
6. Pools — 7. Fins — 8. Esperanto.

(1E6; 5B2b; 7B1a; 7B18b Application 4)

408.

Champion map of Wilmington, New Castle
County, Del. -- Scale [ca. 1:24,000]. --
[Daytona Beach, Fla.] : Champion Map Corp. ;
Bensalem, Pa. : Champion Map of Philadelphia
[distributor], c1980.
  1 map : col. ; 101 x 143 cm.
  Shows radial distances.
  Accompanied by: Champion wall map index of
Wilmington, New Castle Co., Del. (10 p. ; 29 cm.).
  Insets: Champion map of New Castle County,
Del. — South New Castle County, Del.
  Copr. no.: VA 47-012.

(4E1; 5E1c; 7B11; 7B19)

409.

Potomac River basin : Clarke, Frederick, and
Highland Counties, Virginia, Berkeley, Grant,
Hampshire, Hardy, Jefferson, Mineral, Morgan,
and Pendleton Counties, West Virginia :
outstanding water storage sites / U.S.
Department of Agriculture, Soil Conservation
Service, [and] Economics, Statistics, and
Cooperatives Service, [and] Forest Service, and
W.V. Department of Natural Resources. -- Scale
1:562,000 (W 79°40'--W 77°40'/N 39°40'--
N 38°20'). -- Lanham, MD : USDA-SCS, 1980.
  1 map : col. ; 27 x 32 cm.
  Shows reservoirs.
  "October 1978."
  "Source: USDA Soil Conservation Service."
  Includes location map.
  "14,512."

(1E1; 4C3)

410.

Tax map, Galloway Township, Atlantic County,
N.J. / E.M. Wootton, Aug. 1931. -- Rev. / by
Edward M. Wootton ... Oct. 1st, 1933. -- Scale
[ca. 1:24,000]. -- [Mays Landing] : Township
Engineer, [1933]
  2 maps on 2 sheets : photocopy ; 61 x 92 cm.
  Shows block numbers and sheet lines of large
scale tax maps.
  Includes cachet of approval by "State Tax
Department."
  Contents: Key sheet north — Key sheet south.

(2C1)

411.

Kutztown State College, campus map / campus
map drawn by Andrew Szoke. -- Not drawn to
scale. -- [Kutztown? Pa. : s.n., 1980?]
  1 view : col. ; on sheet 28 x 43 cm., folded
to 22 x 14 cm.
  Panel title: Welcome to Kutztown State
College : campus map, office directory,
transportation guide, area map.
  Oriented with north toward the lower left.
  Bird's-eye view.
  Indexed.
Text, directory, 2 location maps, and ill. on
verso.

(3B7; 5B1a Application; 5D1h; 7B4; 7B10a)

412.

Mount St. Helens and vicinity (Wash.-Oreg.)
1:100,000-scale topographic map : April-1980 /
United States Department of the Interior,
Geological Survey. -- Special ed. -- Scale
1:100,000. "1 centimeter on the map represents
1 kilometer on the ground" ; universal
transverse Mercator proj. (W 123°00'00"-- W
121°52'30"/N 40°37'30"--N 45°52'30"). --
Reston, Va. : The Survey, 1980.
  1 map : col. ; 83 x 86 cm., folded to
25 x 10 cm.
  Relief shown by contours and spot heights.
  Panel title.
  Title from lower right margin: Mount St.
Helens and vicinity, Wash.-Oreg.
  "Produced by the U.S. Geological Survey from
1:100,000-scale quadrangles."
  "Road update furnished by Washington State
Department of Natural Resources."
  "Projection and 10,000-meter grid, zone 10,
universal transverse Mercator 25,000-foot grid
ticks based on Washington coordinate system,
south zone, and Oregon coordinate system, north
zone."

(1B8b Application 1; 3B2; 3C1; 3D1b Application;
7B8)

413.

Coronado National Forest, Arizona and New
Mexico : Chiricahua, Peloncillo, Dragoon
mountain ranges / U.S. Department of
Agriculture, Forest Service, Southwestern
Region ; compiled and drafted at Regional
Office, Albuquerque, N.M. -- Rev. 1975, repr.
1980. -- Scale [1:126,720]. ½" = 1 mile ;
polyconic proj. (W 110°15'--W 108°45'/N 32°10'
-- N 30°19'). -- Albuquerque : The Office,
[1980]
  2 maps on 1 sheet : both sides, col. ;
80 x 64 cm. and 51 x 63 cm., sheet 82 x 65 cm.
  "Forest Service map class A."
  Relief shown by hachures and spot heights.
  "Gila and Salt River meridian, New Mexico
principal meridian."
  Includes location map, recreation indexes,
"Vicinity map and U.S.G.S. index," col. ill.,
and text.
  Contents: Douglas Ranger District, Coronado
National Forest (Chiricahua-Peloncillo Mts.)
Arizona and New Mexico — Douglas Ranger
District, Coronado National Forest (Dragoon
Mts.) Arizona.

(5D1b Application; 7B1a Application)

414.

Fresno & vicinity wall map / Thomas Bros.
Maps. -- Scale [ca. 1:31,000]. -- Los
Angeles : Thomas Bros., [1978]
  1 map : col. ; 128 x 173 cm.
  Insets: Firebaugh — Mendota — San
Joaquin — Kingsburg — Orange Cove — Fresno
& vicinity arterial map — Kerman — Coalinga
— Huron.
  "60448 B278P378."
  Copr. no.: VA 56-640.

(1F3; 7B18b Application 5; 7B19)

415.
Gravity field of the northwest Pacific Ocean basin and its margin, Kuril Island arc-trench system / compiled by Anthony B. Watts, Mikhail G. Kogan, John H. Bodine, 1977 ; drawn by Portia Takakjian. -- Scale [ca. 1:2,520,000]. "At 55°N" (W 135°--W 165°/N 60°--N 35°). -- Boulder, Colo. : Geological Society of America, c1978 (Washington, D.C. : William & Heintz Map Corp.)
    1 map : col. ; 92 x 77 cm., folded in envelope 31 x 23 cm. -- (Map and chart series / Geological Society of America ; MC-27)
    "Contour interval 25mgal."
    Data sources: Lamont-Doherty Geological Observatory ... [et al.]
    "Lamont-Doherty Geological Observatory contribution no. 2726."
    Accompanied by text. (4 p. : ill. ; 28 cm.)
    Includes location map.
    Bibliography in accompanying text.
    Copr. no.: VA 26-846.

(3B2; 4G4; 6E1; 7B11)

416.
Mount Rainier National Park / National Park Service, U.S. Department of the Interior. -- Scale 1:72,000. -- [Washington] : The Service, 1981 printing.
    1 map : col. ; 47 x 61 cm., folded to 24 x 11 cm.
    Relief shown by shading and spot heights.
    Panel title: Mount Rainier.
    "Reprint 1981."
    Text and ill. (some col.) on verso.

(3B1a; 4F6; 5D1h; 7B4; 7B7a)

417.
Iraq, synopsis 1980. -- Scale 1:2,000,000 (E 38°--E 51°/N 38°--N 27°). -- Geneva, Switzerland : Petroconsultants, [1981]
    1 map : photocopy ; 63 x 61 cm.
    Shows oil and gas leases, fields, and wells.
    "January 1981."
    At head of title: Foreign Scouting Service.
    Includes "Geologic sketch map," location map, "List of rightholders," and "Summary of activity during 1980."

(1F2)

418.
The Oxford map of Qatar / published by GEOprojects in association with Oxford University Press. -- Scale 1:270,000 (E 50°30'--E 51°45'/N 26°30'--N 24°30'). -- Beirut, Lebanon : GEOprojects, c1980 (Mitcham, Surrey : Cook, Hammond, and Kell)
    1 map : col. ; 75 x 41 cm., folded to 22 x 13 cm.
    Relief shown by gradient tints and spot heights.
    Panel title: The Oxford map of Qatar : with city map of Doha.
    Includes index and location map.

Text, col. ill., indexed map of Doha, and "Businessman's guide" on verso.

(4G4)

419.
Oman regions. -- Scale [ca. 1:6,000,000] (E 51°--E 60°/N 27°--N 16°). -- [Washington] : Central Intelligence Agency, [1980]
    1 map : col. ; 21 x 17 cm.
    "Unclassified."
    "504408 1-80 (544486)."

(1F2; 7B14)

420.
Project map, Spring Creek Watershed, Colbert and Franklin counties, Alabama / U.S. Department of Agriculture, Soil Conservation Service. -- Scale [ca. 1:90,000] ; transverse Mercator proj. (W 87°44'--W 87°30'/N 34°45'--N 34°30'). -- Fort Worth, Tex. : USDA-SCS, 1979.
    1 map : col. ; 35 x 25 cm.
    Shows floodwater retarding structure, benefited area, channel work, drainage area controlled by structure, and grade control structure.
    "Source: Data compiled by Watershed Planning Staff."
    "Base compiled from USGS quadrangle sheets ... "
    Includes location map.
    "April 1979 4-R-36445. September 1976 base 4-R-35682."

(3C1; 7B1a; 7B7a)

421.
The peoples of China ; The People's Republic of China / produced by the Cartographic Division, National Geographic Society ; Richard J. Darley, chief cartographer ; John E. Shupe, associate chief cartographer. -- Scale 1:7,150,000. 1 cm. = 71.5 km. or 1 in. = 113 miles (E 74°--E 134°/N 54°--N 19°). -- Scale 1:6,000,000. 1 cm. = 60 km. or 1 in. = 94.7 miles. ; Albers conical equal-area proj., standard parallels 24° and 48° (E 66°--E 142°/N 54°--N 14°). -- Washington, D.C. : National Geographic Society, 1980.
    2 maps on 1 sheet : both sides, col. ; 76 x 94 cm. and 74 x 92 cm., sheet 78 x 96 cm., folded to 16 x 24 cm.
    Relief of "The People's Republic of China" shown by shading and spot heights. Depths shown by contours and soundings.
    English and romanized Chinese (Pinyin).
    "Supplement to the National Geographic, July 1980, page 2A, vol. 158, no. 1-China."
    Includes text, "A key to ethnolinguistic groups in China," col. portraits of ethnic groups, col. ill., "Pronunciation guide to Pinyin," and list of "Geographical equivalents."

(1G3; 3B Application; 5B2b; 5C1 Application)

422.
Drummond Island. -- Scale [ca. 1:45,000]. --
[Saginaw, Mich.] : Jon W. Ledy, c1979.
1 map : col. ; 52 x 87 cm.
Relief shown by contours.

(4D3; 4F6)

423.
Ground-water resources of Harrison County /
by Katie Crowell ; cartography, Craig A.
Schottenstein. -- Scale [ca. 1:63,000]. --
Columbus : Ohio Dept. of Natural Resources,
Division of Water, 1980.
1 map : col. ; 53 x 84 cm.
Relief shown by contours and spot heights.
Includes location map.

(1F1; 4D1)

424.
Main road map of Cuyahoga County / compiled,
published, and copyrighted by Commercial Survey
Co. -- Scale [ca. 1:250,000]. -- Cleveland,
Ohio : The Company, c1979.
1 map ; 19 x 26 cm.
Inset: Main road map of downtown Cleveland.
Copr. no.: VA 47-634.

(1F1; 4D5)

425.
Missouri railroad map / Department of
Transportation, State of Missouri. -- Scale
[1:950,000]. 1 in. represents approx. 15
miles. -- [Jefferson City] : The Department,
1979.
1 map : col. ; on sheet 45 x 58 cm.
Indexed.

(3B2; 4D5; 5D1e)

426.
Presettlement vegetation of Kalamazoo
County, Michigan / by Thomas H. Hodler ... [et
al.]. -- Scale [ca. 1:63,000]. -- [Kalamazoo] :
Western Michigan University, c1981.
1 map : col. ; 62 x 62 cm., on sheet
64 x 97 cm.
"Funded by Lucia Harrison Endowment Fund,
Department of Geography, W.M.U."
Includes text.

(1F5; 4F6; 7B6)

427.
Viaduct clearances for Chicago streets and
neighboring suburbs / M & E Enterprise ; art &
color by Debb, 1980. -- Scale [1:42,451]. 1
in. equals approx. 0.67 miles. -- Bolingbrook,
Ill. : M & E Enterprise, c1980.
1 map : both sides, col. ; 110 x 67 cm., on
sheet 61 x 89 cm.

Variant title: Viaduct clearances.
Includes text and indexes.

(1B8b Application 1; 1F1; 3B1b; 3B2; 4F6;
5D1j; 7B4)

428.
The city of Crystal Lake, vicinity. -- Scale
[ca. 1:19,500]. -- [Woodstock, Ill.] : John E.
Bailey, c1979.
1 map : photocopy, col. ; 51 x 49 cm.
"Final field inspection, May 5, 1979."
"ColorReplica maps."
Includes illegible indexes.

(7B1a Application; 7B12)

429.
Hennepin County, Minnesota / U.S. Department
of Agriculture, Soil Conservation Service [and]
United States Department of the Interior,
Geological Survey. -- Scale [ca. 1:125,000] ;
universal transverse Mercator proj. (W 93°47'30"
--W 93°09'00"/N 45°16'00"--N 44°46'00"). --
[Lincoln, Neb.] : The Service, [1979?]
1 map : col. ; 44 x 40 cm.
"Base source: USGS 1:100,000 county base
compiled in 1976."
Includes location map.
"5,P-37,508."

(3C1; 3D1b; 4C3; 4D5; 4F7; 7B7a)

430.
Ludington, Mich. : 1892 / drawn and
published by C.J. Pauli. -- Not drawn to
scale. -- Milwaukee, Wis. : C.J. Pauli,
[1892]
1 view : photocopy ; 50 x 89 cm.
Bird's-eye view.
Indexed.

(3B7; 5B1a Application; 5C1 Application;
7B10a)

431.
Wetzler, Plan = Wetzlar, map / Städte-Verlag
E. v. Wagner & J. Mitterhuber. -- 10. Aufl. --
Scale 1:12,000. -- Stuttgart-Bad Cannstatt :
Städte-Verlag, [1980?]
1 map : col. ; 58 x 47 cm., folded to
20 x 13 cm.
Relief shown by spot heights.
Title in German, English, and French.
Panel title.
Includes indexes, inset of city center, and
map of Wetzlar/Giessen region.

(1B5; 4C1; 5D1h; 7B2; 7B3)

432.
Kābul, Afghanistan city graphic 1:15,000 /
prepared and published by the Defense Mapping
Agency Hydrographic/Topographic Center. -- Ed.

2-DMA. -- Scale ca. 1:15,000 (E 69°07'00"--
E 60°14'31"/N 34°34'32"--N 34°27'15"). --
Washington : The Center, [1981]
    1 map : col. ; 89 x 79 cm. -- (Series /
Defense Mapping Agency, Hydrographic/
Topographic Center ; U911)
    "Compiled in 1980 from best available source
materials."
    Relief shown by contours and spot heights.
    Legend and place names in English and Pushto.
    "Grid--Atlas."
    "Distribution limited--destroy when no longer
needed."
    Includes indexes to streets and points of
interest, glossary, and boundary diagram.
    "Stock no. U911XKABUL."

(1B10; 2B1; 3B1d; 6E1)

(On map: Approximate scale 1:15,000)

450.
    Rhône-Alpes : le portrait d'une région,
itinéraires pour une découverte : 140 cartes au
1/100 000 / mises au point pour cet ouvrage par
l'Institut géographique national ; préf. de
Pierre Doueil ; présentation de Roger Frison-
Roche et Jacques Soustelle ; textes d'André
Lugagne. -- Éd. 1. -- Scale 1:100,000. --
Paris : l'Institut, c1978.
    1 atlas (160 p.) : col. ill., col. maps ; 28
cm. -- (Livrecartes)
    Cover title: La région Rhône-Alpes.
    Includes index.
    Bibliography: p. 159.
    ISBN 2-258-00392-X : 80.00F

(1F6; 8D1)

451.
    Fishery atlas of the northwest African shelf :
volumes I and II = Atlas rybacki szelfu Afryki
północno-zachodniej : tom I i II / Andrzej
Klimaj ; translated from Polish [by B.
Przybylska]. -- Scales differ (W 21°--W 5°/
N 36°--N 9°). -- Warsaw, Poland : Published for
the National Marine Fisheries Service, U.S.
Dept. of Commerce, and the National Science
Foundation, Washington, D.C., by the Foreign
Scientific Publications Dept. of the National
Center for Scientific, Technical, and
Economic Information ; Springfield, Va. :
available from the U.S. Dept. of Commerce,
National Technical Information Service, 1976.
    1 atlas (2 v. in 1 (221 p.)): ill., maps ;
30 cm.
    Translation of: Atlas rybacki szelfu
Afryki północno-zachnodniej / Andrzej Klimaj.
v. 1-1971, v. 2-1973.
    Bibliography: p. 219-[220]
    TT73-54097.

(1D1; 1F1; 4C3; 4D4; 5B22; 7B2 Application;
7B19)

452.
    The Grosset world atlas. -- Scales differ. --
New York : Grosset & Dunlap, c1977.
    1 atlas (48 p.) : col. ill., col. maps ;
29 cm.
    "Entire contents copyright 1977 by Hammond
Incorporated"—Verso t.p.
    Includes index.
    ISBN 0-448-13156-0 (lib. ed.)

(4F6)

453.
    Suomen kansankulttuurin kartasto = Atlas of
Finnish folk culture / toimittanut Toivo
Vuorela. -- Scale [ca. 1:4,000,000 and
1:8,000,000]. -- Helsinki : Suomalaisen
Kirjallisuuden Serua, 1976-
    1 atlas (v.) : col. maps ; 35 cm. --
(Suomalaisen Kirjallisuuden Seuran
toimituksia, ISSN 0355-1768 ; 325)
    Finnish and German with captions also in
English.
    Transparent overlay attached to last fly
leaf: v. 1.
    Bibliography: v. 1, p. 12, 150-151.
    Contents: 1. Aineellinen kulttuuri —

    ISBN 951-717-099-8 (v. 1)

(1D2; 3B1a Application 2; 5B20 Application;
5C3; 6B1; 6F1; 7B2)

454.
    The Oxford atlas / edited by Sir Clinton
Lewis [and] J.D. Campbell ; with the
assistance of D.P. Bickmore and K.F. Cook. --
1st ed. repr. with revision to date, 1967. --
Scales differ. -- London : Oxford University
Press, 1967.
    1 atlas (96, xxvi, 93 p.) : col. maps ;
40 cm.
    Includes gazetteer and index.

(1F4; 2B1; 5B5)

455.
    Atlante delle regioni d'Italio / Istituto
geografico Visceglia-Roma. -- Scale 1:250,000.
1 cm. = 2500 m. -- Roma : Istituto geografico
Visceglia, [197-?]-
    1 atlas (portfolios) : folded maps ;
35 cm.
    Cover title.

(3B2; 4F7)

456.
Society in view : a graphic atlas for the social sciences / general editor, J.A. Johnston ; editor, Col Cunnington ; designer, John Braben ; chief cartographer, John Feodoroff. -- Scales differ. -- Milton, Qld. : Jacaranda Press, 1978.
1 atlas (vi, 121 p.) : col. ill., col. maps ; 29 cm.
Includes indexes.
Contents: Individual, family and society — Communities — Australian settlement — National themes — Spaceship earth — The future.
ISBN 0-7016-0974-5 (pbk.)

(1F6; 7B18b Application 4; 8E1)

457.
Portugal, atlas do ambiente. -- Scale 1:1,000,000 ; Gauss proj. (W 9°30'--W 6°25'/ N 42°10'--N 36°50'). -- [Lisbon?] : Presidência do conselho de Ministros, Secretaria de Estado do Ambiente, Comissão Nacional de Ambiente, [1975]-
1 atlas (1 v. (loose-leaf)) : col. maps ; 75 cm.
Title from map plate 1.0.
International ellipsoid.
Contents: Ambiente físico — Ambiente biológico — Ambiente biofísico —

(3A2; 5B12; 7B3; 7B8)

458.
Atlas of the Mid-Ocean Dynamics Experiment (MODE-1) / by the MODE-1 Atlas Group ; editors, Valery Lee, Carl Wunche, board of editors, N.P. Fofonoff ... [et al.]. -- Scales differ (W 82°--W 59°/N 35°--N 20°). -- Cambridge, Mass. : Massachusetts Institute of Technology, 1977.
1 atlas (274 p.) : ill., maps ; 23 x 31 cm.
One folded map in pocket.
Bibliography: p. 250-273.

(1F5; 5C2c)

459.
Guide de la route : France, Belgique, Suisse / [réalisé par Sélection du Reader's Digest en collaboration avec le Touring Club de France ; la cartographie routière originale à l'échelle du 1/500 000 a été dressée par Kümmerly et Frey]. -- Ed. 1. -- Scale 1:500,000. -- Paris : Sélection du Reader's Digest, c1969.
1 atlas (455 p.) : ill. (some col.), col. maps ; 29 cm.
Inserted: "Les grands itinéraires" (1 col. map, scale [ca. 1:2,200,000], c1970) and plasticized legend with guide to contents.
Includes index.

(1F1; 1F6; 7B11; 7B18b Application 6)

460.
Die Erde : Meyers Grosskarten-Edition : zum 150jährigen Bestehen der Kartographie im Bibliographischen Institut / herausgegeben vom Geographisch-Kartographischen Institut Meyer unter Leitung vom Adolf Hanle. -- Scales differ. -- Mannheim : Bibliographisches Institut, 1978.
1 atlas (2 v.) : 72 col. maps (chiefly folded) ; 52 cm.
Scale of most maps 1:5,000,000 and 1:25,000,000.
Issued in case.
Includes index.
Contents: [1] Die Karten — [2] Das Register.
ISBN 3-411-01742-2 (Lw. in Schuber). -- ISBN 3-411-01745-7 (Ldr. in Schuber)

(1E2; 5B20; 8E1)

461.
Phytoplankton production : atlas of the International Indian Ocean Expedition / by J. Krey, B. Babenerd. -- Scales differ ; equal area proj. (W 20°--W 150°/N 30°--N 45°). -- Kiel : Institut für Meereskunde an der Universität Kiel ; Paris : available from Intergovernmental Oceanographic Commission, 1976.
1 atlas (70 leaves) : ill., col. maps ; 42 cm.
"Data used cover a total of twenty years: 1951-1971"—Intro.
"Supported by the Intergovernmental Oceanographic Commission of Unesco, Deutsche Forschungsgemeinschaft".
Principal maps at scale ca. 1:34,000,000.

(4D4; 7B6)

462.
Africa, Erdteilatlas für Blinde : mit 22 Reliefkarten / von Paul Georgi. -- Scales differ. -- Leipzig : Deutsche Zentralbücherei für Blinde zu Leipzig, [1974-1975]
1 atlas (22 leaves of braille and tactile graphics) : maps ; 44 cm. + text (xiii, 95 p. of braille ; 36 cm.)
Braille t.p.
Includes loose, rolled sheets of braille text in atlas v.
Includes Blindenschrift-Alphabet, p. [2] of cover in text.
Contents: [1] Kartenband — [2] Erlauterungsband.
"Bestellnummer 151."

(5B26; 5E1d; 7B19)

463.
[Braille world atlas / produced by Ministry of Education]. -- Scales differ. -- [Riyadh, Saudi Arabia : The Ministry, 1966?]
1 atlas ([18] leaves of braille and tactile graphics) : ill., maps ; 30 cm.
Ms. cover.

Arabic braille with brief English ms.
captions.
Lacking t.p.

(0B7; 1B7; 5B26)

464.
Worcester book of maps for travel training. --
Scales differ. -- [Worcester, Worcestershire :
W.J. Pickles, 1967]
1 atlas ([11] leaves of braille and tactile
graphics) : maps ; 30 cm.
Cover title.
Braille on cover: Embossed maps of Worcester.
Attributed to W.J. Pickles by: Joseph W.
Wiedel, donor.
Cover stamped: Worcester College for the
Blind.

(4C3; 5B26)

465.
[Combination atlas map of Miami County,
Indiana / compiled, drawn, and published from
personal examinations and surveys by Kingman
Brothers]. -- Scales differ. -- [Chicago? :
Kingman Brothers, 1877]
1 atlas (103 p.) : ill., hand col. maps,
ports. ; 45 cm.
Title from reprint: Knightstown, Ind. :
Bookmark, 1974.
History of Miami County / John A. Graham: p.
[13]-29.
Business directories: p. [99]-103.
LC copy imperfect: t.p. wanting.

(0B4; 1B12; 5C3 Application; 7B3; 7B18b; 7B20a)

466.
Atlas of Licking Co., Ohio, combination atlas,
and 1875 atlas. -- Scales differ. --
Knightstown, Ind. : Mayhill Publications, 1970.
1 atlas (32 p.) : ill., maps ; 44 cm.
Reprint. Originally published: Atlas of
Licking Co., Ohio. New York : Beers, Soule &
Co., 1866.
Reprint. Originally published: Combination
atlas map of Licking County, Ohio. L.H. Everts,
1875.

(1B1a Application; 7B7; 11A)

467.
Historical atlas. -- Scales differ. --
Maplewood, N.J. : Hammond Incorporated, [1980?]
1 atlas (H-48, U-64, 32 p.) : col. ill., col.
maps ; 32 cm.
At head of title: United States Air Force
Academy.
Cover title.
Contents: World history atlas — United
States history atlas — Hammond headline world
atlas.

(5B5; 7B6)

*(Collective title stands alone; separate t.p.
precedes each of the three works listed in
"Contents" note.)*

468.
Southam easy-fold maps of Montreal and 34
municipalities. -- 1st ed. -- Scale ca.
1:18,000. 1500 ft. to 1 in. [and] ca.
1:24,000. 2000 ft. to 1 in. -- Montreal,
Quebec : Southam Printing Co., Montreal
Division, Map Dept., c1960.
1 atlas (78 p.) : ill., folded maps (some
col.) ; 21 cm.
"Printed and published annually."
"Places of interest and a brief history of
Montreal"—p. 22-26.
"Includes index."

(3B2; 3B4; 5C2d; 7B18b)

*(Scale on maps reads, e.g. "Scale: app.
1:18,000 ...")*

*(This is the 1st ed. of a work to be "printed
and published annually"; but at the time of
cataloguing, six years after publication, no
subsequent issue had been received so the atlas
was treated as a monograph.)*

469.
The map abstract of trends in calls for police
service, Birmingham, Alabama, 1975-1976 /
Raymond O. Sumrall, Jane Roberts, John P.
Zakanycz ; Neal G. Lineback. -- Scale
[ca. 1:140,000]. -- University, Ala. :
University of Alabama Press, c1978.
1 atlas (iv, 93, [4] p.) : maps ; 22 x 28
cm. -- (Map abstract / Dept. of Geology and
Geography, University of Alabama ; no. 5)
"A cooperative effort between the Birmingham
Police Department ... the School of Social Work
and the Department of Geology and Geography of
The University of Alabama"—Pref.
LC copy has alternate pages printed upside
down.
ISBN 0-8173-9006-5.

(1F6; 4F6; 5D2; 6E1; 7B6; 7B20a)

470.
Peru-Chile, boundary dispute maps, 1544-1879 :
photographs of 300 maps. -- Scales differ. --
[S.l. : s.n., 192-?]
1 atlas (1 case (300 leaves of plates)) : 300
maps ; 61 cm.
Title from label on case.
A collection of photocopies of maps from
various sources.

(1B1a; 1E1; 3B6; 4C6; 4D7; 4F7; 5B21; 5C2a; 7B3;
7B6; 7B10a)

471.

Geografía postal de España : especial para oposiciones a ejecutivos y auxiliares de correos / Gonzalo García Sánchez. -- Scales differ. -- [Madrid? : s.n.], 1977 (Madrid : Litograph)

1 atlas (65 leaves) : 64 col. maps ; 24 x 34 cm.

Cover title.

ISBN 8440041683.

(4C6; 4D7; 4G1; 5C2; 5D2)

472.

Atlante stradale d'Italia : Sud : Campania, Puglia, Basilicata, Calabria, Sicilia : scala 1:200 000 / Touring club italiano. -- 1980, 4^a ed. rinnovata. -- Scale 1:200,000. -- Milano : T.C.I., 1980.

1 atlas ([13], 55, [15] p.) : 55 maps (chiefly col.) ; 33 cm.

Index map and distance chart on front lining paper.

Includes index.

(1F3; 2B1; 3B1a; 4D5; 5B5; 5B6)

(Publisher's name transcribed from imprint directly; this is a common initialism for the organization.)

473.

Metsker's atlas of Benton County, State of Washington : dated July 1976. -- Scale [ca. 1:31,680]. 2 in. - 1 mile. -- Tacoma, Wash. : Metsker Maps, 1976.

1 atlas (50 [i.e. 66] leaves) : photocopy, all maps ; 39 x 46 cm.

"Township maps show property owners".

Caption title.

On each map: Copyright by Thos. C. Metsker.

Blue line print.

Includes index.

(1B2; 3B2; 5B7; 7B3; 7B6; 7B9; 7B10a)

474.

Guía urbana de Barcelona. -- 17^a ed. general, 1^a ed., 1976. -- Scales differ. -- Barcelona : José Pamias Ruiz, 1976.

1 atlas (2 v.) : col. maps (some folded) ; 16 cm.

Two folded col. maps in jacket pocket (21 x 9 cm.)

Includes indexes.

Contents: t. 1. Guía urbana de Barcelona — t. 2. Entidad municipal metropolitana y otros municipios.

ISBN 84-85212-04-5 (complete work).

(2D2; 5C2f; 7B18b Application 4; 8B2)

475.

Maps of India, 1795-1935 / Maharashtra State Archives, Govt. of Maharashtra ; [editor, B.C.

Kunte, associates, V.T. Gondil, A.K. Kharade]. -- Scales differ. -- [Bombay] : The Archives, 1978 [i.e. 1979]

1 atlas (ix, 31 p. (1 folded)) : 23 maps (some col.) ; 42 cm.

Maps relating to the then Bombay Presidency reproduced from the collection in the Directorate of Archives.

Preface date 1979.

Rs20.00

(1F3; 1F6; 4D5; 4F2; 5C2a; 7B1a Application)

476.

Index, lands subject to investigation for areas now or formerly below mean high water / prepared for the Natural Resource Council by State of New Jersey, Department of Environmental Protection, Office of Environmental Analysis. -- Scale [ca. 1:48,000]. -- Trenton, N.J. : The Office, 1979.

1 atlas ([150] leaves) : chiefly maps ; 36 cm.

"Photography and index by Mark Hurd Aerial Surveys, Inc."—On map sheets.

Errata sheet inserted.

(1F4; 4D5; 7B6; 7B18b)

477.

Journeys of Frodo : an atlas of J.R.R. Tolkien's The lord of the rings / Barbara Strachey. -- 1st Ballantine Books ed., Mar. 1981. -- Scales differ. -- New York, N.Y. : Ballantine Books, 1981.

1 atlas ([109] p.) : 51 col. maps ; 19 x 25 cm.

"Contour lines ... are 50, 100 or 200 feet apart according to the scale"—Foreword.

"At various scales, from 100 miles to the inch to 1/4 mile to the inch"—p. [1]

Datum for the map grid is Hobbiton Hill.

ISBN 0-345-29723-7. -- ISBN 0-345-29633-8 (pbk).

(2B1; 5B6; 5D2; 7B1a Application; 7B8; 8B2)

478.

CIMA, Channel Industries Mutual Aid / [specially prepared for Channel Industries Mutual Aid by Map Graphics Inc.]. -- Scale 1:31,680. 1 in. = 1/2 mile. -- Houston, Tex. : Map Graphics, [c1978]

1 atlas ([58] p., 1 folded leaf of plates) : col. maps ; 30 cm.

Locations of selected industrial sites in Houston, Tex.

Cover title.

Includes directory: p. [2]-[9]

(1B1a; 3B2; 5B13; 5B14; 7B1a; 7B3; 7B18b)

(Title from cover; first page begins with statement of responsibility and continues with publication data.)

Appendix **H**

Abbreviations

H.1 Use abbreviations in descriptive catalogue records as instructed below and in *AACR2* Appendix B. (B.1 mod.)

H.2 Use the following categories of abbreviations in the title and statement of responsibility area and the statement of responsibility element in the edition area; also use them in titles and statements of responsibility in the series area and in contents notes:

  a) those found in the prescribed sources of information for the particular area
  b) *i.e., et al.*, and their equivalents in nonroman scripts (cf. 0F, 1F5)

(B.4)

H.3 Use abbreviations elsewhere in the catalogue entry, subject to the limitations specified in footnotes to section H.7 below. Do not use them if the brevity of the statement makes them unnecessary or if the resulting statement might not be clear. Do not use single letter abbreviations to begin a note. Do not abbreviate words in quoted notes.

(B.5)

H.4 Use an abbreviation for the corresponding word in another language if the abbreviation commonly used in that language has the same spelling. In case of doubt, do not use the abbreviation. (B.6)

H.5 Use a prescribed abbreviation for the last part of a compound word, e.g., Textausg. for Textausgabe. (B.7)

H.6 In inflected languages, use the abbreviation of a word given in the following lists in the nominative case for an inflected form of that word. If, however, the abbreviation includes the final letter(s) of the word, modify the abbreviation to show the final letter(s) of the inflected form, e.g., литература, лит-ра; литературы, лит-ры.

(B.8)

H.7 GENERAL ABBREVIATIONS   (B.9 mod.)

TERM	ABBREVIATION	TERM	ABBREVIATION
årgang	årg.	and others	et al.
aastakäik	aastak	Anno Domini	A.D.
Abdruck	Abdr.	approximately	approx.
abgedruckt	abgedr.	argraffiad	arg.
Abteilung, Abtheilung	Abt.	átdolgozott	átdolg.
afdeling	afd.	Auflage	Aufl.
aflevering	afl.	augmenté, -e	augm.
altitude	alt.[1]	augmented	augm.
and	&[2]	aumentada	aum.

1. Use only in recording mathematical data in entries for cartographic materials.
2. Use only in uniform titles in listing languages.

TERM	ABBREVIATION	TERM	ABBREVIATION
aumentato	aum.	died	d.
Ausgabe	Ausg.	diena	d.
avdeling	avd.	djilid	djil.
Bändchen	Bdchn.	document	doc.
Bände	Bde.	dopunjeno	dop.
Band	Bd.	drukarnia	druk.
Before Christ	B.C.	edition, -s	ed., eds.
binary coded decimal	BCD	édition	éd.
bind	bd.	editor	ed.[6]
black and white	b&w	enlarged	enl.
bogtrykkeri	bogtr.	equinox	eq.[5]
boktrykkeri	boktr.	ergänzt	erg.
book	bk.	erweitert	erw.
born	b.	establecimiento	
bövített	böv.	tipográfico	estab. tip.
broj	br.	et alii	et al.
Brother, -s	Bro., Bros.[3]	et cetera	etc.
Buchdrucker,		évfolyam	évf.
Buchdruckerei	Buchdr.	facsimile, -s	facsim.,
Buchhandlung	Buchh.		facsims.
bulletin	bull.	fascicle	fasc.
bytes per inch	bpi	fascicule	fasc.
capitolo	cap.	flourished	fl.
část	č.	folio	fol.
centimetre, -s	cm	following	ff.
century	cent.[4]	foot, feet	ft.
cetekan	cet.	frame, -s	fr.
chapter	ch.	fratelli	f.lli[3]
circa	ca.	Gebrüder	Gebr.[3]
číslo	čís.	gedruckt	gedr.
colored, coloured	col.	genealogical	geneal.
Compagnia	Cia.	godina	g.
Compagnie	Cie	government	govt.
Compañía	Cía.	Government Printing	
Company	Co.	Office	G.P.O.
compare	cf.	Handschrift, -en	Hs., Hss.
compiler	comp.	Her (His) Majesty's	
confer	cf.	Stationery Office	H.M.S.O.
copyright	c	Hermanos	Hnos.[3]
Corporation	Corp.	hour, -s	hr.
corrected	corr.	id est	i.e.
corretto, -a	corr.	Idus	Id.
corrigé	corr.	illustration, -s	ill.
část	cz.	illustrator	ill.[6]
declination	decl.[5]	imienia	im.
deel	d.	imprenta	impr.
del (Danish, Norwegian,		imprimerie	impr.
and Swedish)	d.	inch, -es	in.
département	dép.	including	incl.
Department	Dept.	Incorporated	Inc.[3]
diameter	diam.	introduction	introd.

---

3. Use only in names of firms and other corporate bodies.
4. Use in headings and in indicating the period when a manuscript was probably written.
5. Use only in recording mathematical data in entries for cartographic materials.
6. Use only in a heading as a designation of function (see *AACR2* 21.0D).

TERM	ABBREVIATION	TERM	ABBREVIATION
izdája	izd.	paperback	pbk.
izmenjeno	izm.	part, -s	pt., pts.
jaargang	jaarg.	parte	pt.
Jahrgang	Jahrg.	partie, -s	ptie, pties
javitott	jav.	photograph, -s	photo., photos.
jilid	jil.	plate number	pl. no.
Kalendae	Kal.	poprawione	popr.
kiadás	kiad.	portrait, -s	port., ports.
kilometre, -s	km	posthumous	posth.
kniha	kn.	predelan	pred.
knjiga	knj.	preface	pref.
kötet	köt.	preliminary	prelim.
księgarnia	księg.	printing	print.[8]
leto	l.	privately printed	priv. print.
librairie	libr.	projection	proj.[9]
Lieferung	Lfg.	proširen	proš.
Limited	Ltd.[7]	przekład	przekł.
livraison	livr.	przerobione	przerob.
maatschappij	mij.	pseudonym	pseud.
manuscript, -s	ms., mss.	publishing	pub.
ménuo	mėn.	redakcja	red.
metai	m.	refondu, -e	ref.
miesięcznik	mies.	réimpression	réimpr.
millimetre, -s	mm	report	rept.
minute, -s	min.	reprinted	repr.
miscellaneous	misc.	reproduced	reprod.
Nachfolger	Nachf.[7]	reviderade	revid.
nakład	nakł.	revisé, -e	rev.
nakladatelství	nakl.	revised	rev.
naukowy	nauk.	revu, -e	rev.
neue Folge	n. F.	right ascension, -s	RA[9]
new series	new ser.	riveduto	riv.
New Testament	N.T.	ročník	roč.
no name (of publisher)	s.n.	rocznik	rocz.
no place (of publication)	s.l.	rok	r.
Nonae	Non.	rozszerzone	rozsz.
nouveau, nouvelle	nouv.	second, -s	sec.
number, -s	no.	série	sér.
numbered	numb.	series	ser.
numer	nr.	sešit	seš.
numero (Finnish)	n:o	signature	sig.
numéro (French)	no	sine loco	s.l.
numero (Italian)	n.	sine nomine	s.n.
número (Spanish)	no.	skład główny	skł. gł.
Nummer	Nr.	stabilimento tipográfico	stab. tip.
nummer	nr.	številka	št.
nuovamente	nuov.	stronica	str.
odbitka	odb.	superintendent	supt.
oddział	oddział	Superintendent of	
Old Testament	O.T.	Documents	Supt. of Docs.
omarbeidet	omarb.	supplement	suppl.
oplag	opl.	svazek	sv.
opplag	oppl.	szám	sz.
opracowane	oprac.	tahun	th.
otisk	ot.	talleres gráficos	tall. gráf.
page, -s	p.	Teil, Theil	T.

7. Use only in names of firms and other corporate bodies.
8. Do not use in recording the date of printing in the publication, distribution, etc., area (cf. 4F6, *AACR2* 2.4G2).
9. Use only in recording mathematical data in entries for cartographic materials.

TERM	ABBREVIATION	TERM	ABBREVIATION
tipografía, tipográfica	tip.	upplaga	uppl.
tiskárna	tisk.	utarbeidet	utarb.
title page	t.p.	utgave	utg.
tjetakan	tjet.	utgiven	utg.
tome	t.	uzupełnione	uzup.
tomo	t.	verbesserte	verb.
towarzystwo	tow.	vermehrte	verm.
translator	tr.	volume, -s	v., vol.,[10] vols.[10]
typographical	typog.	vuosikerta	vuosik.
typographie, typographique	typ.	vydáni	vyd.
udarbejdet	udarb.	wydanie	wyd.
udgave	udg.	wydawnictwo	wydawn.
udgivet	udg.	wydział	wydz.
uitgaaf	uit.	založba	zal.
uitgegeven	uitg.	zeszyt	zesz.
uitgevers	uitg.	zväzok	zv.
umgearbeitet	umgearb.	zvezek	zv.
Universitäts-Buchdrucker, Universitäts-Buchdruckerei	Univ.-Buchdr.		

10. Use at the beginning of a statement and before a roman numeral.

☐ POLICIES

British Library

The symbols km, cm, and mm are used for metric measurements.

Library of Congress

The Library of Congress uses km., cm., and mm. for metric measurements.

National Library of Australia

The National Library of Australia uses km., cm., and mm. for metric measurements.

National Library of New Zealand

The symbols km., cm., and mm. are used for metric measurements.

National Map Collection, PAC

The metric symbols for kilometre, centimetre, and millimetre are km, cm, and mm (without full stops) as prescribed by the International System of Units (SI). In accordance with Canadian Government Policy, the National Map Collection uses all metric symbols according to the SI system, without full stops except at the end of a sentence.

H.8 ABBREVIATIONS TO BE USED IN CITING BIBLIOGRAPHIC SOURCES OF DATA

Use common, self-explanatory abbreviations of the type listed below in citing the source of data used in the catalogue entry, so long as the use of abbreviations does not obscure the language of the source cited.

TERM	ABBREVIATION	TERM	ABBREVIATION
American	Amer.	dictionary	dict.
annuaire	ann.	directory	direct.
annuario	ann.	encyclopedia	encycl.
anuario	an.	English	Engl.
bibliography	bibl.	history	hist.
biography	biog.	Katalog	Kat.
British	Brit.	literature	lit.
catalog, catalogue	cat.	littérature	litt.
cyclopedia	cycl.	museum	mus.
diccionario	dicc.	national	nat.
		report	rept.

(B.13)

H.9 ABBREVIATIONS OF NAMES OF CERTAIN COUNTRIES, THE STATES OF AUSTRALIA AND THE UNITED STATES, THE PROVINCES OF CANADA, AND THE TERRITORIES OF AUSTRALIA, CANADA AND THE UNITED STATES

Use the following abbreviations of place names, other than the names of cities and towns, as additions to certain other place names (cf. *AACR2* 23.4), as additions to names of certain corporate bodies (cf. *AACR2* 24.4C), as additions to the name of the place of publication or distribution, in the place of publication, distribution, etc., area (cf. 4C3), and in notes. Do not abbreviate names omitted from the list.   (B.14)

□ POLICIES
    Library of Congress
        Use the abbreviations of state names recommended by the U.S. Postal Service only if they appear in the chief source of information. The abbreviations for state names in H.9 are still valid; use them either when supplying the name of the state or abbreviating it when the full form appears.
        Transcribe the U.S. Postal Service abbreviations as they appear, whether in upper case, or in both upper and lower case, with or without full stops, i.e., Ca, Ca., CA, CA.

TERM	ABBREVIATION	TERM	ABBREVIATION
Alabama	Ala.	Manitoba	Man.
Alberta	Alta.	Maryland	Md.
Arizona	Ariz.	Massachusetts	Mass.
Arkansas	Ark.	Michigan	Mich.
Australian Capital		Minnesota	Minn.
Territory	A.C.T.	Mississippi	Miss.
British Columbia	B.C.	Missouri	Mo.
California	Calif.	Montana	Mont.
Colorado	Colo.	Nebraska	Neb.
Connecticut	Conn.	Nevada	Nev.
Delaware	Del.	New Brunswick	N.B.
District of Columbia	D.C.	New Hampshire	N.H.
Distrito Federal	D.F.	New Jersey	N.J.
Florida	Fla.	New Mexico	N.M.
Georgia	Ga.	New South Wales	N.S.W.
Illinois	Ill.	New York	N.Y.
Indiana	Ind.	New Zealand	N.Z.
Kansas	Kan.	Newfoundland	Nfld.
Kentucky	Ky.	North Carolina	N.C.
Louisiana	La.	North Dakota	N.D.
Maine	Me.	Northern Territory	N.T.

TERM	ABBREVIATION	TERM	ABBREVIATION
Northwest Territories	N.W.T.	Tennessee	Tenn.
Nova Scotia	N.S.	Territory of Hawaii	T.H.
Oklahoma	Okla.	Texas	Tex.
Ontario	Ont.	Union of Soviet Socialist	
Oregon	Or.	Republics	U.S.S.R.
Pennsylvania	Pa.	United Kingdom	U.K.
Prince Edward Island	P.E.I.	United States	U.S.
Puerto Rico	P.R.	Vermont	Vt.
Queensland	Qld.	Victoria	Vic.
Rhode Island	R.I.	Virgin Islands	V.I.
Russian Soviet		Virginia	Va.
Federated Socialist		Washington	Wash.
Republic	R.S.F.S.R.	West Virginia	W. Va.
Saskatchewan	Sask.	Western Australia	W.A.
South Australia	S. Aust.	Wisconsin	Wis.
South Carolina	S.C.	Wyoming	Wyo.
South Dakota	S.D.	Yukon Territory	Yukon
Tasmania	Tas.		

(B.14)

H.10 ABBREVIATIONS OF THE NAMES OF THE MONTHS   (B.15)

BELORUSSIAN	BULGARIAN	CZECH	DANISH
студз.	ян.	led.	jan.
лют.	февр.	ún.	febr.
сак.	март	břez.	marts
крас.	април	dub.	april
май	май	květ.	maj
чэрв.	юни	červ.	juni
ліп.	юли	červen.	juli
жнівень	авг.	srp.	aug.
верас.	септ.	září	sept.
кастр.	окт.	říj.	okt.
ліст.	ноем.	list.	nov.
снеж.	дек.	pros.	dec.

DUTCH	ENGLISH	ESTONIAN	FRENCH
jan.	Jan.	jaan.	janv.
feb.	Feb.	veebr.	févr.
maart	Mar.	märts	mars
apr.	Apr.	apr.	avril
mei	May	mai	mai
juni	June	juuni	juin
juli	July	juuli	juil.
aug.	Aug.	aug.	août
sept.	Sept.	sept.	sept.
oct.	Oct.	okt.	oct.
nov.	Nov.	nov.	nov.
dec.	Dec.	dets.	déc.

GERMAN	GREEK, MODERN	HUNGARIAN	INDONESIAN AND MALAYSIAN
Jan. (Jän.)	'Ιαν.	jan.	Jan. (Djan.)
Feb.	Φεβρ.	feb.	Peb.
März	Μάρτ.	márc.	Mrt.
Apr.	'Απρ.	ápr.	Apr.
Mai	Μάιος	máj.	Mei (Mai)
Juni	'Ιούν.	jun.	Juni (Djuni)
Juli	'Ιούλ.	jul.	Juli (Djuli)
Aug.	Αὔγ.	aug.	Ag.

GERMAN	GREEK, MODERN	HUNGARIAN	INDONESIAN AND MALAYSIAN
Sept.	Σεπτ.	szept.	Sept.
Okt.	Ὀκτ.	okt.	Okt.
Nov.	Νοέμ.	nov.	Nop.
Dez.	Δεκ.	dec.	Des.

ITALIAN	LATIN	LATVIAN	LITHUANIAN
genn.	Ian.	jan.	saus.
febbr.	Febr.	feb.	vas.
mar.	Mart.	marts	kovas
apr.	Apr.	apr.	bal.
magg.	Mai.	maijs	geg.
giugno	Iun.	junijs	birž.
luglio	Iul.	julijs	liepa
ag.	Aug.	aug.	rugp.
sett.	Sept.	sept.	rugs.
ott.	Oct.	okt.	spalis
nov.	Nov.	nov.	lapkr.
dic.	Dec.	dec.	gr.

NORWEGIAN	POLISH	PORTUGUESE	ROMANIAN
jan.	stycz.	jan.	Ian.
febr.	luty	fev.	Feb.
mars	mar.	março	Mar.
april	kwiec.	abril	Apr.
mai	maj	maio	Mai
juni	czerw.	junho	Iunie
juli	lip.	julho	Iulie
aug.	sierp.	agosto	Aug.
sept.	wrzes.	set.	Sept.
okt.	paźdz.	out.	Oct.
nov.	listop.	nov.	Noem.
des.	grudz.	dez.	Dec.

RUSSIAN	SERBO-CROATIAN		SLOVAK	
янв.	jaн.	siječ.	l'ad.	jan.
февр.	фебр.	velj.	ún.	feb.
март (мартъ)	март	ožuj.	brez.	mar.
апр.	април	trav.	dub.	apr.
май	мaj	svib.	kvet.	máj.
июнь (iюнь)	jyни	lip.	ćerv.	jún
июль (iюль)	jyли	srp.	ćerven.	júl
авг.	ауг.	kol.	srp.	aug.
сент.	септ.	ruj.	zári.	sept.
окт.	окт.	list.	ruj.	okt.
ноябрь	нов.	stud.	list.	nov.
дек.	дец.	pros.	pros.	dec.

SLOVENIAN	SPANISH	SWEDISH	UKRAINIAN	WELSH
jan.	enero	jan.	січ.	Ion.
feb.	feb.	febr.	лют.	Chwe.
mar.	marzo	mars	бер.	Mawr.
apr.	abr.	april	квіт.	Ebr.
maj	mayo	maj	трав.	Mai
jun.	jun.	juni	чер.	Meh.
jul.	jul.	juli	лип.	Gorf.
avg.	agosto	aug.	серп.	Awst
sept.	sept. (set.)	sept.	вер.	Medi
okt.	oct.	okt.	жовт.	Hyd.
nov.	nov.	nov.	лист.	Tach.
dec.	dic.	dec.	груд.	Rhag.

219

Abbreviations of the names of the months in languages not listed above may be used if they appear in style manuals for the language concerned.　(B.15)

Appendix **J**

Numerals

J.0 INTRODUCTORY RULE

Apply the following rules to all items published after 1820. Apply them also to items published before 1821 unless rules 1B12–16, 2B8–9, 4B9, 7B10, 7B20 instruct otherwise.　(C.0)

J.1 ARABIC VS. ROMAN　(C.1)

J.1A　Retain the roman numeral in headings for persons identified by roman numerals (e.g., rulers, popes) and in headings for corporate bodies whose names include roman numerals unless, in the case of a corporate body, a particular rule instructs otherwise (cf. *AACR2* 24.7B2).　(C.1A)

> John XXIII, *Pope*

> XXth Century Heating & Ventilating Co.

J.1B　Retain roman numerals in uniform titles that are integral parts of the name of the work. In the case of numerals used to identify particular parts of a work, follow instructions of the appropriate rule (cf. *AACR2* 25.6A2, 25.18A3).　(C.1B)

> Sancho II y el cerco de Zamora

J.1C　Substitute arabic numerals for roman in the following areas of the bibliographic description:

1) in the edition or issue statement in the edition area
2) in the material (or type of publication) specific details area unless a particular rule directs otherwise (cf. 3B2) or unless J.1D applies
3) in the date of publication, distribution, etc., in the publication, distribution, etc., area (cf. 4F1)

> , 1971
> *not*　, MCMLXXI

4) in the other physical details element of the physical description area
5) in the series numbering in the series area unless J.1D applies.　(C.1C)

J.1D　Retain roman numerals if the substitution of arabic numerals makes the statement less clear; for example, when roman and arabic numerals are used in conjunction to distinguish the volume, section, series, or other group from the number, part, or other division of that group.　(C.1D)

(The Washington papers ; vol. IV, 36)

J.1E   When roman numerals are retained, write them in capitals except those used in paging or page references and those appearing in lowercase in the chief source of information or in quoted notes. Use lowercase roman numerals in paging or page references even though capitals appear in the item.   (C.1E)

xliii, 289 p.

J.2   ARABIC VS. NUMERALS SPELLED OUT   (C.2)

J.2A   Retain spelled-out numerals in the names of corporate bodies unless a particular rule instructs otherwise (cf. *AACR2* 24.7B2).   (C.2A)

Four Corners Geological Society

J.2B   Retain spelled-out numerals in uniform titles that are integral parts of the name of the work. In the case of numerals used to identify particular parts of a work, follow the instructions of the appropriate rule (cf. *AACR2* 25.6A2, 25.18A3).

(C.2B)

Quinze joies de mariage

J.2C   Substitute arabic numerals for spelled-out numerals in the following areas of the bibliographic description:

1)   in the edition or issue statement in the edition area
2)   in the material (or type of publication) specific details area unless a particular rule directs otherwise (cf. 3B2)
3)   in the date of publication, distribution, etc., element in the publication, distribution, etc., area, unless a particular rule directs otherwise (cf. 4F1)
4)   in the physical description area
5)   in the series numbering in the series area.   (C.2C)

J.3   NUMERALS BEGINNING NOTES

Spell out a numeral that is the first word of a note that is not quoted.   (C.3)

First ed. published in 1954

Four no. a year, 1931; 5 no. a year, 1932–1934

J.4   ORIENTAL NUMERALS   (C.4)

J.4A   Substitute roman numerals or Western-style arabic numerals for numerals in the vernacular in the cataloguing of Arabic alphabet, Far Eastern, Greek, Hebrew, Indic, etc., materials as instructed in the following rules.   (C.4A)

J.4B   Use roman numerals in romanized headings for persons identified by numerals (e.g., rulers).   (C.4B)

J.4C Substitute Western-style arabic numerals for vernacular numerals in romanized headings for corporate bodies and in uniform titles. (C.4C)

> Thawrat 25 Māyū, 1969

> al-Lajnah al-ʿUlyā li-Iḥtifālāt 14 Tammūz

J.4D Substitute Western-style arabic numerals in the following areas of the bibliographic description:

1) in the edition or issue statement element in the edition area
2) in the material (or type of publication) specific details area unless a particular rule directs otherwise (cf. 3B2)
3) in the date of publication, distribution, etc., element of the publication, distribution, etc., area
4) in the physical description area
5) in the series numbering in the series area. (C.4D)

J.4E Consider inclusive dates and other numbers to be a single unit in languages that are read from right to left, e.g., 1960–1965, not 1965–1960. Add punctuation following the unit at the left, e.g., .1973–1976. (C.4E)

J.5 INCLUSIVE NUMBERS (C.5)

J.5A Give inclusive numbers in full, e.g., p. 117–128, 1967–1972. (C.5A)

J.6 ALTERNATIVE DATES

When alternative dates of birth or death are given in headings for persons (cf. *AACR2* 22.18), word the second of the alternatives as it is spoken, e.g., 1506 *or* 7; 1819 *or* 20; 1899 *or* 1900. In all other cases, repeat all of the digits common to the two numbers, e.g., 1971 *or* 1972. (C.6)

J.7 ORDINAL NUMERALS (C.7)

J.7A In the case of English-language items, record ordinal numerals in the form 1st, 2nd, 3rd, 4th, etc. (C.7A)

J.7B In the case of other languages, follow the usage of the language if ascertainable,[1] e.g., French, 1er, 1ère, 2e, 3e, etc.; German, 1., 2., 3., etc.; Italian, 1°, 1^a, 2°, 2^a, 3°, 3^a, etc. (C.7B)

J.7C For Chinese, Japanese, and Korean items, accompany the arabic numeral by the character indicating that the numeral is ordinal. (C.7C)

J.7D If the usage of a language cannot be ascertained, use the form 1., 2., 3., etc. (C.7D)

---

1. The following publication is a useful source for the form of ordinal numerals in European languages:

Allen, C.G. *A Manual of European Languages for Librarians* / C.G. Allen. — London ; New York : Bowker, 1975.

GLOSSARY

Abbreviations in parentheses following the definitions indicate their sources; see end of glossary for bibliographic citations.

Access point. A name, term, code, etc., under which a bibliographic record may be searched and identified.
(*AACR2*)

Added entry. An entry, additional to the main entry, by which an item is represented in a catalogue; a secondary entry. *See also* Main entry. (*AACR2*)

Added title page. A title page preceding or following the title page chosen as the basis for the description of the item. It may be more general, as a series title page, or equally general, as a title page in another language. (*AACR2*)

Aerial chart. *See* Aeronautical chart.

Aerial mosaic. *See* Uncontrolled photomosaic.

Aerial photograph. *See* Air photograph.

Aerial remote-sensing image. An image produced by a remote-sensing device located in an aircraft.

Aeronautical chart. A map designed to assist navigation in the air. (I.C.A.)

Air photo. *See* Air photograph.

Air photograph. Any photograph taken from the air. When used in a cartographic context, this term normally refers to photographs of the surface of the earth (or other celestial body) taken downwards, vertically, or at a predetermined angle from the vertical. *See also* High oblique air photograph, Low oblique air photograph, Orthophotograph, Vertical air photograph. (I.C.A.)

Alternative title. The second part of a title proper that consists of two parts, each of which is a title; the parts are joined by the word *or* or its equivalent in another language, e.g., The tempest, or, The enchanted island.
(*AACR2*)

Altitude tint. *See* Hypsometric tint.

Anaglyphic map. A map specially printed in two colours in such a way that when viewed through twin eye-pieces of the same two colours, a three-dimensional impression of relief is seen. (I.C.A.)

Analytical entry. An entry for a part of an item for which a comprehensive entry has been made.
(*AACR2*)

Analytical note. The statement in an analytical entry relating the part being analyzed to the comprehensive work of which it is a part. (*AACR2*)

Anamorphic map. A map which is characterized by an areal distortion in proportion to a factor other than a linear one.

Ancillary map. 1. A small supplementary or secondary map outside the neat line of the principal or main map.
2. A generic term for small supplementary or secondary maps located either inside or outside the neat line of the principal or main map.
See also Inset.

Glossary

Anonymous. Of unknown authorship. *(AACR2)*

Architectural rendering. A pictorial representation of a building intended to show, before it has been built, how the building will look when completed. *(AACR2)*

Area. 1. A major section of the bibliographic description, comprising data of a particular category or set of categories. *See also* Geographic area. *(AACR2)*

Atlas. 1. A volume of maps, plates, engravings, tables, etc., with or without descriptive letterpress. It may be an independent publication or it may have been issued to accompany one or more volumes of text.

(AACR2)

    2. A collection of maps designed to be kept (bound or loose) in a volume. *(I.C.A.)*

    See also Appendix F.

Atlas factice. A composite work made up of a selection of previously issued maps. It may be bound or loose-leaf.

Author. *See* Personal author.

Author-title added entry. *See* Name-title added entry.

Author-title reference. *See* Name-title reference.

Back-to-back. This term is used in the physical description area for either (i) two maps being two versions, in two languages, of one work, printed tête-bêche on one sheet; or (ii) the two parts of a bilingual atlas published tête-bêche in one volume. *See also* Tête-bêche.

Bar scale. A subdivided line which shows distances at a given scale. *See also* Graphic scale. *(I.C.A.)*

Base map. 1. A map used as a framework on which to depict thematic concepts and/or other information.

    2. A map used as a primary source for compilation. *(I.C.A.)*

Bathymetric tint. A colour applied to the area between selected depth contours to depict the physical relief of the floor of the ocean, or the bed of another body of water. *See* illustration in 7B1. *See also* Hypsometric tint. *(Boggs)*

Bibliographic description. A set of bibliographic data recording and identifying a publication, excluding access points; i.e., the description that begins with the title proper and ends with the last note in the note area.

Binder's title. The title lettered on the cover of an item by a binder, as distinguished from the title on the publisher's original cover. *See also* Spine title. *(AACR2)*

Bird's-eye view. A perspective representation of the landscape, as it might be visible from a high view-point above the surface of the earth, or other celestial body, in which detail is shown as if projected on to an oblique plane. *(I.C.A.)*

Bleed. *See* Bleeding edge.

Bleeding edge. An edge of a map to which printed detail extends after the paper has been trimmed. Also known as bleeding margin format, bleed, extended colour (USA). *(I.C.A.)*

Bleeding margin format. *See* Bleeding edge.

Block diagram. A representation of the landscape in either perspective or isometric projection, usually with some vertical exaggeration. Block diagrams may also be used to illustrate subterranean structures.

(I.C.A.)

Blueline. A cartographic item having blue lines with a white background. Bluelines are made from vellum, film positives, or any translucent or transparent original on which an image has been made and reproduced by the whiteprint process. (Not the same as a *Blueprint:* the whiteprint process is the opposite of the dark-print process from which blueprints are made.) *(Stevenson)*

Blueline print. *See* Blueline.

Blueprint. A photographic print, white on a bright blue background, made on paper, cloth, etc., sensitized with potassium ferricyanide and a ferric salt, and developed, after exposure, by washing in plain water. *See also* Facsimile reproduction. *(Boggs)*

Border. The area of a map which lies between the neat line and the outermost line of the surrounding frame. *(I.C.A.)*

Braille map. A map utilizing symbols and the braille system of writing for tactile use by the blind.

Cadastral map. A map which delineates property boundaries. *See also* Plat. *(I.C.A.)*

224

Cartogram. 1. A small diagram on the face of a map displaying quantitative data.

2. Synonym for diagrammatic map. (I.C.A.)

Cartographic material. Any material representing, in whole or in part, the earth or any celestial body at any scale; includes two- and three-dimensional maps and plans; aeronautical, navigational, and celestial charts; globes; block diagrams; sections; aerial, satellite, and space photographs; atlases; bird's-eye views, etc.

(*AACR2*)

Cartouche. A panel on a map, often with decoration, enclosing the title or other legends, the scale, etc.

(Stamp)

Catalogue. 1. A list of library materials contained in a collection, a library, or a group of libraries, arranged according to some definite plan.

2. In a wider sense, a list of materials prepared for a particular purpose, e.g., an exhibition catalogue, a sales catalogue. (*AACR2*)

Celestial chart. A map representing the heavens. (Based on: Stamp)

Celestial globe. *See* Globe.

Celestial sphere. An imaginary sphere of infinite radius, described about an assumed centre, and upon which imagined positions of celestial bodies are projected along radii passing through the bodies. For observations on bodies within the limits of the solar system, the assumed centre is the centre of the earth. For bodies where the parallax is negligible, the assumed centre may be the point of observation.

(U.S. Defense Mapping Agency)

Chart. 1. A map designed primarily for navigation. (I.C.A.)

2. In cartography, a special-purpose map generally designed for the use of navigators (e.g., an aeronautical chart, a nautical chart), although the word is also used to designate other types of special purpose maps, e.g., a celestial chart (i.e., a "star chart"). *See also* Hydrographic chart. (*AACR2*)

Chart index. *See* Index map.

Chief source of information. The source of bibliographic data to be given first preference as the source from which a bibliographic description (or portion thereof) is prepared. (*AACR2*)

Collaborator. One who works with one or more associates to produce a work; all may make the same kind of contribution, as in the case of shared responsibility (*see AACR2* 21.6), or they may make different kinds of contributions, as in the case of collaboration between an artist and a writer (*see AACR2* 21.24). *See also* Joint author, Mixed responsibility, Shared responsibility. (*AACR2*)

Collection. 1. If by one author: three or more independent works or parts of works published together. If by more than one author: two or more independent works or parts of works published together and not written for the same occasion or for the publication in hand. (*AACR2*)

2. An artificial accumulation of manuscripts or documents devoted to a single theme, person, event, or type of record. (Evans)

3. A body of manuscripts or papers, including associated printed or near-print materials, having a common source. If formed by or around an individual or family, such materials are more properly termed *personal papers* or *records*. If the cumulation is that of a corporate entity, it is more properly termed *records*. (Evans)

Collective title. A title proper that is an inclusive title for an item containing several works. *See also* Uniform title (2). (*AACR2*)

Colophon. 1. A statement at the end of an item giving information about one or more of the following: the title, author(s), publisher, printer, date of publication or printing; it may include other information.

(*AACR2*)

2. The term *colophon* refers also to an emblem or design identifying a printer, publisher, or cartographer; it may appear within the content portion of the cartographic item or as marginal information.

(Stevenson)

Coloured illustration. An illustration in two or more colours. (*AACR2*)

Compiled map. A map incorporating information collected from various sources; not compiled from survey data made for the map in question. (U.S. Defense Mapping Agency)

Glossary

Compiler. One who produces a collection by selecting and putting together matter from the works of various persons or bodies. Also, one who selects and puts together in one publication matter from the works of one person or body. *See also* Editor. *(AACR2)*

Component. A component is one of two or more maps all of equal importance for the purpose of bibliographic identification or access appearing on a single sheet. No map is the dominant or "main" map, and they may or may not have a collective title and/or individual titles. *See also* Part, Segment.

Composite atlas. *See* Atlas factice.

Compound surname. A surname consisting of two or more proper names, often connected by a hyphen, conjunction, or preposition. *(AACR2)*

Contour. A line joining points of equal vertical distance above or below a datum. Such a line on a map is a type of isoline. *N.B.:* This term, contour, is to be distinguished from outline (a line, or lines, bounding a discrete area on the face of a map) for which many languages use a cognate word. *See* illustration in 7B1. *(I.C.A.)*

Contour interval. The vertical distance (on the ground) between two successive contours. *(I.C.A.)*

Contour map. A map which portrays relief by the use of contour lines. (U.S. Defense Mapping Agency mod.)

Controlled photomosaic. An assembly of parts of vertical air photographs joined together to leave minimal scale variations. In a controlled photomosaic, the distortions of perspective have been adjusted to ground measurements. *See also* Orthophotomosiac. *(I.C.A.)*

Conventional title. *See* Uniform title.

Coordinates, geodetic. *See* Geodetic coordinates.

Coordinates, geographic. *See* Geographic coordinates.

Corporate body. An organization or group of persons that is identified by a particular name and that acts, or may act, as an entity. Typical examples of corporate bodies are associations, institutions, business firms, nonprofit enterprises, governments, government agencies, religious bodies, local churches, and conferences. *(AACR2)*

Cover title. A title printed on the original cover of an item. *(AACR2)*

Cross-reference. *See* Reference.

Declination. 1. In a system of polar or spherical coordinates, the angle at the origin between a line to a point and the equatorial plane, measured in a plane perpendicular to the equatorial plane.

2. The arc between the equator and the point measured on a great circle perpendicular to the equator.

3. (astronomy) The angular distance to a body on the celestial sphere measured north or south through 90° from the celestial equator along the hour circle of the body. Comparable to latitude on the terrestrial sphere. (U.S. Defense Mapping Agency)

Diagram. A graphic representation of numerical data, or of the course or results of an action or process. The term is sometimes applied also to maps characterized by much simplified, or schematic, representation. *(I.C.A.)*

Diazo print. A photograph or photocopy produced on a surface (as paper) by coating with a solution containing a diazo compound that is decomposed on exposure to light, the compound in the unexposed parts being then converted to a coloured image formed by an azo dye by developing, especially with an alkaline solution or gaseous ammonia. *See also* Facsimile reproduction. (Webster's)

Distributor. An agent or agency that has exclusive or shared marketing rights for an item. *(AACR2)*

Double leaf. A leaf of double size with a fold at the fore edge or at the top edge of the book. *(AACR2)*

Dyeline print. *See* Diazo print.

Edition. 1. In the case of books and booklike materials, all those copies of an item produced from substantially the same type image, whether by direct contact or by photographic methods.

2. In the case of nonbook materials, all the copies of an item produced from one master copy and issued by a particular publishing agency or a group of such agencies. Provided the foregoing conditions are fulfilled, a change of identity of the distributing body or bodies does not constitute a change of edition. *See also* Facsimile reproduction, Impression, Issue, Reprint. *(AACR2)*

Editor. One who prepares for publication an item not his own. The editorial labour may be limited to the preparation of the item for the manufacturer, or it may include supervision of the manufacturing, revision (restitution) or elucidation of the text, and the addition of an introduction, notes, and other critical matter. For certain works it may involve the technical direction of a staff of persons engaged in writing or compiling the text. *See also* Compiler. *(AACR2)*

Edge, bleeding. *See* Bleeding edge.

Element. A word, phrase, or group of characters representing a distinct unit of bibliographic information and forming part of an area (q.v.) of the description. *(AACR2)*

Elevation drawing. A drawing which represents phenomena as if projected horizontally on a vertical surface. *N.B.:* In English this term is used mainly in architecture and engineering rather than in cartography where it is not to be confused with elevation. *(I.C.A.)*

Elevation tint. *See* Hypsometric tint.

Ellipsoid. A mathematically defined surface. *(I.C.A.)*

Engraving. *See* Map engraving.

Entry. A record of an item in a catalogue. *See also* Heading. *(AACR2)*

Entry, added. *See* Added entry.

Entry, main. *See* Main entry.

Entry word. The word by which an entry is primarily arranged in the catalogue, usually the first word (other than an article) of the heading. *See also* Heading. *(AACR2)*

Epoch. An arbitrary moment in time to which measurements of position for a body or orientation for an orbit are referred. (U.S. Defense Mapping Agency)

Equinoctial point. *See* Equinox.

Equinox. One of the two points of intersection of the ecliptic and the celestial equator, occupied by the sun when its declination is 0°. Also called equinoctial point. (U.S. Defense Mapping Agency)

Equivalent scale. *See* Verbal scale.

Explanatory reference. An elaborated *see* or *see also* reference that explains the circumstances under which the headings involved should be consulted. *(AACR2)*

Explicit. A statement at the end of the text of a manuscript, or early printed book, or at the end of one of its divisions, indicating its conclusion and sometimes giving the author's name and the title of the work. *(AACR2)*

Extended colour. *See* Bleeding edge.

Facsimile map. A printed reproduction of a map identical with the original. (I.C.A. mod.)

Facsimile reprint. *See* Facsimile reproduction.

Facsimile reproduction. A reproduction that has as its chief purpose to simulate the physical appearance of the original work as well as to provide an exact replica of the text. *See also* Blueprint, Diazo print. *(AACR2)*

Fascicle. One of the temporary divisions of a work that, for convenience in printing or publication, is issued in small installments, usually incomplete in themselves; they do not necessarily coincide with any formal division of the work into parts, etc. Usually the fascicle is protected by temporary paper wrappers and may or may not be numbered. A fascicle is distinguished from a part (q.v.) by being a temporary division of a work rather than a formal component unit. *(AACR2)*

Fence diagram. A drawing in perspective of three or more geologic sections with their relationships to one another. (Gary)

Filing title. *See* Uniform title.

Format. In its widest sense, any particular physical presentation of an item. *(AACR2)*

Form lines. Lines, resembling contour lines, drawn to present a conception of the shape of the terrain without regard to a true vertical datum or regular spacing. *See* illustrations in 7B1. (Thompson, M.)

Gazetteer. A list of geographic names, together with references to their positions and, sometimes, descriptive information. (I.C.A.)

General material designation. A term indicating the broad class of material to which an item belongs, e.g., sound recording. *See also* Specific material designation. *(AACR2)*

Glossary

Geodetic coordinates. A coordinate system, on a specific spherical or spheroidal surface, which expresses the position of points in terms of latitude and longitude. (I.C.A.)

Geographic area. An expanse or tract of the earth's surface. (Webster's)

Geographic coordinates. Coordinates in terms of latitude and longitude. (I.C.A.)

Geographical pole. Either of the two points of intersection of the surface of the earth with its axis, where all meridians meet. (U.S. Defense Mapping Agency)

Globe. The model of a celestial body, usually the earth or a celestial sphere, depicted on the surface of a sphere. (AACR2)

Gradient tint. *See* Hypsometric tint.

Graphic scale. A drawing, or diagram, which enables quantitative measurements to be read off the face of a map, e.g., linear scale; diagonal scale; variable scale. *See also* Bar scale, Slope diagram.

(I.C.A.)

Graticule. 1. An imaginary network of meridians and parallels on the surface of the earth or other celestial body. (a) Division of a circle into 360 degrees; (b) division of a circle into 400 grads.

2. A network of lines, on the face of a map, which represents meridians and parallels. (I.C.A.)

Great circle. A "circle" on the surface of a sphere whose plane passes through the centre and hence bisects it into hemispheres. (I.C.A.)

Greenwich meridian. *See* Prime meridian.

Grid. Network of two sets of uniformly spaced parallel lines, one set intersecting the other at right angles. When superimposed on a map, it usually carries the name of the projection used for the map—that is, Lambert grid, transverse Mercator grid, universal transverse Mercator grid. *See also* Military grid.

Hachures. Short lines, following the direction of maximum slope, which indicate relief. *See* illustration in 7B1. (I.C.A.)

Half title. A brief title of a publication appearing on a leaf preceding the title page. (AACR2)

Heading. A name, word, or phrase placed at the head of a catalogue entry to provide an access point in the catalogue. *See also* Main heading. (AACR2)

High oblique air photograph. An oblique photo which shows the horizon line. (Reeves)

High oblique photograph. A photograph taken with the camera axis intentionally tilted so as to include the apparent horizon. (U.S. Defense Mapping Agency)

Hill shading. Shading employed to create a three-dimensional impression of relief, assuming either vertical or oblique illumination of the ground. In the case of oblique hill shading, the light source is conventionally situated beyond the top left-hand corner of the map. *N.B.:* In English the same term covers the basic concept as well as the means by which it is achieved. *See* illustration in 7B1. *See also* Shaded relief.

(I.C.A.)

Historical map. A map which represents features or phenomena which existed, or which are believed to have existed, in some past period of time. (I.C.A.)

Hydrographic chart. A chart designed to assist navigation at sea or on other waterways. (I.C.A.)

Hypsographic map. *See* Relief map.

Hypsometric layer. A uniform tint or shade covering the area between two successive isolines. (I.C.A.)

Hypsometric map. *See* Relief map.

Hypsometric tint. A colour applied to the area between two selected contours when relief is depicted by a system of layers. *See* illustration in 7B1. *See also* Bathymetric tint. (I.C.A.)

Impression. All those copies of an edition printed at one time. *See also* Issue, Reprint. (AACR2)

Incipit. The opening words of a manuscript or early printed book, or one of its divisions. It frequently includes the word *incipit* or its equivalent in another language. An incipit at the beginning of a work often contains the name of the author and the title of the work. (AACR2)

Incunabulum (map). A very early printed map. *N.B.:* Definition from I.C.A. for which there is no corresponding term in English.

Index map. An index, usually based on an outline map, which shows the layout and numbering system of map sheets which cover an area. (I.C.A.)

Infrared scanning. Scanning using a remote-sensing device which is capable of recording data in the infrared portion of the spectrum.

Initial meridian. *See* Prime meridian.

Inset. A separate map positioned within the neat line of a larger map. *See also* Ancillary map. (I.C.A.)

International meridian. *See* Prime meridian.

International Standard Book Number (ISBN). *See* Standard number.

International Standard Serial Number (ISSN). *See* Standard number.

Isohypse. *See* Contour.

Isoline. A line along which values are, or are assumed to be, constant. (I.C.A.)

Isoline map. A map which represents a continuous distribution by means of isolines. (I.C.A.)

Isometric view. A mode of three-dimensional representation in which distances in all three dimensions are drawn true-to-scale. Lines in the same plane are drawn parallel but arbitrary angles are introduced between planes. (I.C.A.)

Issue. 1. In the case of books and booklike materials, those copies of an edition forming a distinct group that is distinguished from other copies of the edition by more or less slight but well-defined variations; most commonly a new impression for which corrections or revisions have been incorporated into the original type image.

2. In the case of nonbook materials, those copies of an edition of an item forming a distinct group that is distinguished from other copies by well-defined variations. *See also* Impression, Reprint. (*AACR2*)

Item. A document or set of documents in any physical form, published, issued, or treated as an entity, and as such forming the basis for a single bibliographic description. (*AACR2*)

Joint author. A person who collaborates with one or more other persons to produce a work in relation to which the collaborators perform the same function. *See also* Shared responsibility. (*AACR2*)

Key map. A key map is synonymous with an index map. A key map may serve as an index to geographical location depicting the arrangement of sheets that cover a larger area than one of the sheets. (Boggs mod.)

Key-title. The unique name assigned to a serial by the International Serials Data System (ISDS). (*AACR2*)

Landform drawing. A small-scale map showing landforms by the systematic application of a standardized set of simplified pictorial symbols that represent the appearances such forms would have if viewed obliquely from the air at an angle of about 45°. The first major map of this kind was published by Lobeck (1921). *See* illustration in 7B1. (Gary)

Landscape map. A topographic map made to a relatively large scale and showing all details. Such maps are required by architects and landscape gardeners for use in planning buildings to fit the natural topographic features and for landscaping parks, playgrounds, and private estates. These are generally maps of small areas, and a scale is used of 1 in. to 20 ft. to 1 ft. to 50 ft., depending on the amount of detail. (U.S. Defense Mapping Agency)

Latitude. The latitude of a place is its angular distance on a meridian measured northwards or southwards from the terrestrial equator. Hence, on a sphere, it is the angle measured at its centre between the plane of the equator and the radius to any point on the surface of the sphere. *See also* Parallel (of latitude). (I.C.A.)

Layer tint. *See* Hypsometric tint.

Layered (relief) map. A map on which relief is represented by hypsometric layers. *See also* Relief map. (I.C.A.)

Leaf. One of the units into which the original sheet or half sheet of paper, parchment, etc., is folded to form part of a book; each leaf consists of two pages, one on each side, either or both of which may be blank. (*AACR2*)

Linear scale. *See* Bar scale.

Lithography. A method of planigraphic printing invented by Johan Aloysius Senefelder in 1796. To effect a print, the image is drawn, written, or transferred in a greasy medium onto prepared limestone; the non-image areas are dampened while the image is charged with ink which is then transferred to paper. The principle of oil not mixing with water in the alteration of damping and inking is the key to this process in which metal plates have now largely replaced stone. (I.C.A.)

Local meridian. The meridian through any particular place or observer, serving as the reference for local time. (U.S. Defense Mapping Agency)

Glossary

Location map. 1. A map designed to show the position of a particular place, e.g., to show routes of access. 2. A small-scale map inset in, or placed in the margin of, a map at a larger scale to show the location of the area represented by the latter. (I.C.A.)

Longitude. The angle, measured in a plane parallel to that of the equator, between the plane of the meridian through any point and the plane of the Prime meridian or other selected datum meridian. (I.C.A.)

Low oblique air photograph. An oblique photo which does not show the horizon line; the term is restricted by some writers, however, to photos more nearly vertical than horizontal (camera axis less than 45° from vertical). (Reeves)

Low oblique photograph. A photograph taken with the camera axis intentionally tilted so that it does not include the apparent horizon. (U.S. Defense Mapping Agency)

Macroform. A generic term for any medium, transparent or opaque, bearing images large enough to be easily read by the naked eye. *See also* Microform. (*AACR2*)

Main entry. The complete catalogue record of an item, presented in the form by which the entity is to be uniformly identified and cited. The main entry may include the tracings of all other headings under which the record is to be represented in the catalogue. *See also* Added entry. (*AACR2*)

Main heading. The first part of a heading that includes a subheading. (*AACR2*)

Main map. ₍U.S.₎ 1. A map which is augmented by one or more smaller maps, inset or in the margin. (I.C.A.) 2. A primary map selected as the focus for cataloguing.

Manuscript. A writing made by hand (including musical scores), typescripts, and inscriptions on clay tablets, stone, etc. (*AACR2*)

Manuscript map. An original and unprinted map compilation. It may be a map carefully drawn and lettered, or one roughly sketched in either ink or pencil, or a map partially in typewriting; however, it is necessarily a compilation, but, properly, never a hand-drawn tracing of a published map. (Boggs)

Map. A representation, normally to scale and on a flat medium, of a selection of material or abstract features on, or in relation to, the surface of the earth or of another celestial body. *See also* Main map. (*AACR2*)

Map engraving. The process of preparing a map for reproduction by cutting, die-stamping, or etching its features into a plane surface. (I.C.A.)

Map index. *See* Index map.

Map, pictorial. *See* Pictorial map.

Map profile. A scale representation of the intersection of a vertical surface (which may or may not be a plane) with the surface of the ground, or of the intersection of such a vertical surface with that of a conceptual three-dimensional model representing phenomena having a continuous distribution, e.g., rainfall. (I.C.A.)

Map projection. Any systematic arrangement of meridians and parallels, portraying the curved surface of the sphere or spheroid upon a plane. (I.C.A.)

Map section. A scaled representation of a vertical surface (commonly a plane) displaying both the profile where it intersects the surface of the ground or some conceptual model, and the underlying structures along the plane of intersection, e.g., geological section. (I.C.A.)

Map serial. *See* Serial.

Map series. A number of related but physically separate and bibliographically distinct cartographic units intended by the producer(s) or issuing body(ies) to form a single group. For bibliographic treatment, the group is collectively identified by any commonly occurring unifying characteristic or combination of characteristics including a common designation (e.g., collective title, number, or a combination of both); sheet identification system (including successive or chronological numbering systems); scale; publisher; cartographic specifications; uniform format; etc. (The identification of map series is in Appendix D.)

Map view. *See* Bird's-eye view.

Margin. The area of a map sheet which lies outside the border. (I.C.A. mod.)

Marginal data. *See* Marginal information.

Marginal information. Information which appears in the margin of a map. Besides technical data such as explanatory notes, other maps and diagrams, legend, scales, credits, etc., it may also include data used in the bibliographic entry (e.g., title, statement of responsibility, edition, scale, statement of manufacture).

Marginal map. *See* Ancillary map.

Meridian. A great circle arc of 180 degrees terminated by the geographic poles. *See also* Local meridian, Prime meridian. (I.C.A.)

Meridian of origin. *See* Source meridian.

Microfiche. A sheet of film bearing a number of microimages in a two-dimensional array. (*AACR2*)

Microfilm. A length of film bearing a number of microimages in linear array. (*AACR2*)

Microform. A generic term for any medium, transparent or opaque, bearing microimages. *See also* Macroform. (*AACR2*)

Military grid. Two sets of parallel lines intersecting at right angles and forming squares; the grid is superimposed on maps, charts, and other similar representations of the earth's surface in an accurate and consistent manner to permit identification of ground locations with respect to other locations and the computation of direction and distance to other points. (U.S. Defense Mapping Agency)

Mixed authorship. *See* Mixed responsibility.

Mixed responsibility. A work of mixed responsibility is one in which different persons or bodies contribute to its intellectual or artistic content by performing different kinds of activities (e.g., adapting or illustrating a work written by another person). *See also* Shared responsibility. (*AACR2*)

Model. A three-dimensional representation of a real thing, either of the exact size of the original or to scale. (*AACR2*)

Monograph. A nonserial item, i.e., an item either complete in one part or complete, or intended to be completed, in a finite number of separate parts. (*AACR2*)

Moon globe. *See* Globe.

Mosaic. *See* Photomosaic.

Mosaic, controlled. *See* Controlled photomosaic.

Mosaic, uncontrolled. *See* Uncontrolled photomosaic.

Multispectral photo image. An image produced by a remote-sensing device which has utilized two or more spectral bands.

Multispectral scanning image. An image produced by a multispectral scanner (a remote-sensing device) that is capable of recording data in the ultraviolet and visible portions of the spectrum as well as the infrared.

Name-title added entry. An added entry consisting of the name of a person or corporate body and the title of an item. (*AACR2*)

Name-title reference. A reference in which the refer-from line, the refer-to line, or both consist of the name of a person or corporate body and the title of an item. (*AACR2*)

Names. *See* Compound surname, Predominant name.

Natural scale. *See* Representative fraction.

Nautical chart. *See* Hydrographic chart.

Nautical mile. A measure of distance approximately equal to one minute of arc (great circle); International Nautical Mile is equal to 1,852 meters or 6,076.1033 survey feet, or 6,076.11549 U.S. feet. (U.S. Defense Mapping Agency)

Neat line. A line, usually grid or graticule, which encloses the detail of a map. (I.C.A.)

Oblique hill shading. *See* Hill shading.

Oblique photograph. A photograph taken with the camera axis directed intentionally between the horizontal and the vertical. *See also* High oblique air photograph, Low oblique air photograph. (U.S. Defense Mapping Agency)

Orthodrome. *See* Great circle.

Orthophoto. *See* Orthophotograph.

Orthophotograph. A uniform scale air photograph upon which precise horizontal distance measurements can be made. The central perspective of the original photograph is transformed to an orthogonal projection by a process of differential rectification. (I.C.A.)

Orthophotomap. A photomap prepared from a photomosaic of orthophotographs. (I.C.A.)

Orthophotomosaic. A controlled photomosaic assembed from orthophotographs. (I.C.A.)

Other title information. Any title borne by an item other than the title proper or parallel titles; also any phrase appearing in conjunction with the title proper, parallel titles, or other titles, indicative of the character, contents, etc., of the item or the motives for, or occasion of, its production or publication. The term includes subtitles, *avant-titres,* etc., but does not include variations on the title proper (e.g., spine titles, sleeve titles, etc.) *(AACR2)*

Outline map. A map which presents just sufficient geographic information to permit the correlation of additional data placed upon it. (I.C.A.)

Overlay. 1. A transparent sheet containing matter that, when superimposed on another sheet, modifies the data on the latter. *(AACR2)*

2. A print or drawing on a transparent or translucent material, at the same scale as the map to which it is keyed, showing detail that does not appear (or is not emphasized) on the original. (I.C.A.)

Panel title. A title which is printed on a map in such a way that it appears on the outside of the map sheet when the sheet is folded according to a predetermined plan. This title may appear either on the recto or verso of the map. *See also* illustration in 0B3.

Panorama. A perspective representation of the landscape in which the detail is shown as if projected onto a vertical plane or onto the inside of a cylinder centred vertically on the observer. (I.C.A.)

Panoramic drawing. A panorama drawn around a circle, e.g., at a viewing point. (I.C.A.)

Parallel (of latitude). A small circle parallel to the equator, on which all points have the same latitude. (I.C.A.)

Parallel title. The title proper in another language and/or script. *(AACR2)*

Part. 1. One of the subordinate units into which an item has been divided by the author, publisher, or manufacturer. In the case of printed monographs, generally synonymous with *volume* (q.v.); it is distinguished from a fascicle (q.v.) by being a component unit rather than a temporary division of a work.

2. As used in the physical description area, the word *part* designates bibliographic units intended to be bound several to a volume. *(AACR2)*

3. For cartographic material, a part is a physically separate unit that can stand alone bibliographically. It may be used in conjunction with other units of similar design, e.g., a group of maps each of a common area but displaying various themes; or, a group of maps designed primarily for use individually or to provide total coverage of a specific area. *See also* Component, Segment.

Personal author. The person chiefly responsible for the creation of the intellectual or artistic content of a work. *(See AACR2* 21.1A1 for a gloss on this definition.) *(AACR2)*

Perspective view. A representation, in central projection onto a plane surface, of forms (commonly landscape) which exist in three dimensions. (I.C.A.)

Photo image, multispectral. *See* Multispectral photo image.

Photocopy. A macroform photoreproduction produced directly on opaque material by radiant energy through contact or projection. *(AACR2)*

Photograph. *See* Air photograph, High oblique air photograph, High oblique photograph, Low oblique air photograph, Low oblique photograph, Oblique photograph, Orthophotograph.

Photomap. A reproduction of a controlled photomosaic, or of a single rectified air photograph, to which names, symbols, gridlines and/or marginal information have been added. *See also* Orthophotomap. (I.C.A.)

Photomosaic. An assembly of parts of vertical air photographs joined together to leave minimal scale variations. (I.C.A.)

Photomosaic, controlled. *See* Controlled photomosaic.

Photomosaic, uncontrolled. *See* Uncontrolled photomosaic.

Pictorial map. A map in which features are represented by individual pictures in elevation, or perspective, rather than by conventionalized cartographic symbols. (I.C.A. mod.)

Pictorial relief map. A map on which landforms and other topographic features are shown in their correct planimetric position by pictorial symbols representing their appearance from a high oblique view. Not to be confused with bird's-eye view. *See* illustration in 7B1. (I.C.A.)

Plan. 1. A drawing showing relative positions on a horizontal plane, e.g., relative positions of parts of a building, a landscape design; the arrangement of furniture in a room or building; a graphic presentation of a military or naval plan. *(AACR2)*

2. In cartography, a large-scale, detailed map or chart with a minimum of generalization. *(AACR2)*

3. A large-scale, detailed map or chart. Frequently used to describe very large-scale maps in which the outlines of buildings, roads, and other man-made features are shown to scale with little generalization.

(I.C.A.)

4. A map showing a proposed development, e.g., a map illustrating a town planning scheme. (I.C.A.)

Planetable. A field device for plotting the lines of a survey directly from observations. It consists essentially of a drawing board mounted on a tripod, with a leveling device designed as part of the board and tripod.

(U.S. Defense Mapping Agency)

Planetable map. A map compiled by planetable methods. The term includes maps made by complete field mapping on a base projection and field contouring on a planimetric-base map.

(U.S. Defense Mapping Agency)

Planetarium. A model representing the solar system. (Webster's)

Plat. A diagram drawn to scale showing land boundaries and subdivisions, together with all data essential to the description and identification of the several units shown thereon, and including one or more certificates indicating due approval. A plat differs from a map in that it does not necessarily show additional cultural, drainage, and relief features. (U.S. Defense Mapping Agency)

Plate. A leaf containing illustrative matter, with or without explanatory text, that does not form part of either the preliminary or the main sequences of pages or leaves. (AACR2)

Pole, geographical. *See* Geographical pole.

Portfolio. A container for holding loose materials, e.g., maps, paintings, drawings, papers, unbound sections of a book, and similar materials, consisting of two covers joined together at the back; the covers are usually tied with tapes at the fore edge, top, and bottom. (AACR2 mod.)

Predominant name. The name or form of name of a person or corporate body that appears most frequently (1) in the person's works or works issued by the corporate body; or (2) in reference sources, in that order of preference. (AACR2)

Preliminaries. The title page or pages of an item, together with the verso of each title page, any pages preceding the title page(s) and the cover. (AACR2)

Prime meridian. The meridian on the earth's surface from which longitude is measured. Since 1884, the meridian of Greenwich has, by general consent, been recognized as the prime meridian. (I.C.A.)

Print laydown. *See* Uncontrolled photomosaic.

Printing. *See* Facsimile reproduction, Impression, Issue, Lithography, Reprint.

Profile. A scale representation of the intersection of a vertical surface (which may or may not be a plane) with the surface of the ground, or of the intersection of such a vertical surface with that of a conceptual three-dimensional model representing phenomena having a continuous distribution, e.g., rainfall. (I.C.A.)

Projection, map. *See* Map projection.

Radar, sidelooking airborne (SLAR). *See* Sidelooking airborne radar (SLAR) image.

Radar, synthetic aperture (SAR). *See* Synthetic aperture radar (SAR) image.

Recto. 1. The right-hand page of a book, usually bearing an odd page number.

2. The side of a printed sheet intended to be read first. (AACR2)

3. The side of a map sheet intended to be read first, usually bearing the main map(s)/main part of the map (not necessarily bearing the title proper). *See also* Verso.

Reference. A direction from one heading or entry to another. *See also* Explanatory reference. (AACR2)

Reference meridian. *See* Local meridian.

Reference source. Any publication from which authoritative information may be obtained. Not limited to reference works. (AACR2)

Reissue. *See* Issue, Reprint.

Related body. A corporate body that has a relation to another body other than that of hierarchical subordination, e.g., one that is founded but not controlled by another body; one that only receives financial support from another body; one that provides financial and/or other types of assistance to another body, such as "friends" groups; one whose members have also membership in or an association with another body, such as employees' associations and alumni associations. (AACR2)

Glossary

Reliability diagram. A diagram, in the margin of a map, which shows the dates and quality of the source material from which the map has been compiled. (I.C.A.)

Relief. The elevations or the inequalities, collectively, of a land surface; represented on graphics by contours, hypsometric tints, shading, spot elevations, hachures, etc. *See also* Shaded relief.
(U.S. Defense Mapping Agency)

Relief map. A map produced primarily to represent the physical configuration of the landscape, often with hypsometric tints. Not to be confused with plastic relief map or relief model. *See also* Layered (relief) map.
(I.C.A.)

Relief model. A scale representation in three dimensions of a section of the surface of the earth, or other celestial body. A relief model designed to display both physical and cultural features on the surface of the earth is sometimes known as a topographic model. (I.C.A.)

Remote-sensing device. A recording device that is not in physical or intimate contact with the object under study. The remote-sensing technique employs such devices as cameras, lasers, and radio frequency receivers, radar systems, sonar, seismographs, gravimeters, magnetometers, and scintillation counters. *See also* Aerial remote-sensing image, Multispectral photo image, Multispectral scanning image, Sidelooking air-borne radar (SLAR) image, Space remote-sensing image, Synthetic aperture radar (SAR) image, Terrestrial remote-sensing image. (adapted from: Reeves)

Remote-sensing image. An image produced by a remote-sensing device.

Representative fraction. The scale of a map or chart expressed as a fraction or ratio which relates unit distance on the map to distance, measured in the same units, on the ground. (I.C.A.)

Reprint. 1. A new printing of an item made from the original type image, commonly by photographic methods. The printing may reproduce the original exactly (an impression (q.v.)) or it may contain more or less slight but well-defined variations (an issue (q.v.)).

2. A new edition with substantially unchanged text. *See also* Facsimile reproduction. (*AACR2*)

3. A reissue of a map, chart, or atlas either in a state identical with the previous issue (a facsimile reprint) or with limited corrections (a corrected reprint). *See also* Edition. (I.C.A.)

Responsibility, statement of. *See* Statement of responsibility.

Reverse blueline. A cartographic item having a white image or lines on a black or dark-blue background; a print produced by the whiteprint process. *See also* Diazo print, Blueline.

Right ascension. The angular distance measured eastward on the equator from the vernal equinox to the hour circle through the celestial body, from 0 to 24 hours. (U.S. Defense Mapping Agency)

Rock drawing. The stylized representation of steep rock faces which cannot be portrayed by other methods of representing relief. *See* illustration in 7B1. (I.C.A.)

Romanization. Conversion of names or text not written in the roman alphabet to roman-alphabet form.
(*AACR2*)

Running title. The title, or abbreviated title, of the book repeated at the head of each page or at the head of the versos. (*AACR2*)

SAR. *See* Synthetic aperture radar (SAR) image.

SLAR. *See* Sidelooking airborne radar (SLAR) image.

Scale. The ratio of distances on a map, globe, relief model, or (vertical) section to the actual distances they represent. *See also* Graphic scale, Verbal scale, Vertical exaggeration. (I.C.A.)

Schematic map. A map representing features in a much simplified or diagrammatic form. (I.C.A.)

Secondary entry. *See* Added entry.

Section, map. *See* Map section.

Segment. A part of a map where, because of physical limitations, the area being portrayed has been divided to fit on the sheet. *See* Figure 27, 5D1f. *See also* Component, Part.

Serial. A publication in any medium issued in successive parts bearing numerical or chronological designations and intended to be continued indefinitely. Serials include periodicals; newspapers; annuals (reports, yearbooks, etc.); the journals, memoirs, proceedings, transactions, etc., of societies; and numbered monographic series. *See also* Series. (*AACR2*)

Series. A group of separate items related to one another by the fact that each item bears, in addition to its own title proper, a collective title applying to the group as a whole. The individual items may or may not be numbered. (*AACR2*)

234

Series, map. *See* Map series.

Series designation. A series designation is a coded numeric or alphanumeric identification applied to an item (single sheet map, map series, or atlas).

Series number. *See* Series designation.

Shaded relief. A cartographic technique that provides an apparent three-dimensional configuration of the terrain on maps and charts by the use of graded shadows that would be cast by high ground if light were shining from the northwest. Shaded relief is usually used in combination with contours. *See also* Hill shading. *See* illustration in 7B1. (U.S. Defense Mapping Agency)

Shading. *See* Hill shading, Shaded relief.

Shared authorship. *See* Shared responsibility.

Shared responsibility. Collaboration between two or more persons or bodies performing the same kind of activity in the creation of the content of an item. The contribution of each may form a separate and distinct part of the item, or the contribution of each may not be separable from that of the others. *See also* Joint author, Mixed responsibility. (*AACR2*)

Sheet. As used in the physical description area, a single piece of paper other than a broadside, with manuscript or printed matter on one or both sides. (*AACR2*)

Sheet index. *See* Index map.

Sidelooking airborne radar (SLAR) image. An airborne radar, viewing at right angles to the axis of the vehicle, which produces a presentation of terrain or moving targets. (U.S. Defense Mapping Agency)

Sine loco (s.l.). Without place, i.e., without the name of the place of publication. (*AACR2*)

Sine nomine (s.n.). Without name, i.e., without the name of the publisher. (*AACR2*)

Sketch map. A map drawn freehand and greatly simplified which, although preserving general space relationships, does not accurately preserve scale or area. (I.C.A.)

Slide. Transparent material on which there is a two-dimensional image, usually held in a mount, and designed for use in a projector or viewer. (*AACR2*)

Slope diagram. A graphic scale in the margin of a map, from which the angle of slope between contours can be deduced. (I.C.A.)

Sounding. 1. The measured or charted depth of water.

2. A measurement of the depth of water expressed in feet or fathoms and reduced to the tidal datum shown in the chart title. *See* illustration in 7B1. (U.S. Defense Mapping Agency)

Source meridian. The central meridian of a longitude zone of a grid, i.e., a straight line coinciding with the north-south axis of the grid. (I.C.A.)

Space remote-sensing image. An image produced by a remote-sensing device which is located in a spacecraft.

Specific material designation. A term indicating the special class of material (usually the class of physical object) to which an item belongs, e.g., sound disc. *See also* General material designation. (*AACR2*)

Spine title. The title that appears on the spine of an item. *See also* Binder's title. (*AACR2*)

Spot elevation. *See* Spot height.

Spot height. 1. A point on a map which represents the position of an indicated altitude. A spot height is seldom marked on the ground. (I.C.A.)

2. A point on a map or chart whose height above a specified reference datum is noted, usually by a dot or a small sawbuck and elevation value. Elevations are shown, wherever practicable, for road forks and intersections, grade crossings, summits of hills, mountains, and mountain passes, water surfaces of lakes and ponds, stream forks, bottom elevations in depressions, and large flat areas. *See* illustration in 7B1. (U.S. Defense Mapping Agency)

Standard number. The International Standard Number (ISN), e.g., International Standard Book Number (ISBN), International Standard Serial Number (ISSN), or any other internationally agreed upon standard number, that uniquely identifies an item. (*AACR2*)

Standard title. *See* Uniform title.

Star chart. *See* Celestial chart.

State (of maps). An impression that varies from another impression from the same plate because of some change in the plate. The changes in the plate may be: 1) geographic, e.g., deletions, additions, corrections, or alterations in geographical features or names; 2) bibliographic, e.g., changes of title or imprint, recutting of engraved lines, restippling of faded areas, corrections in spelling, additions or deletions to the illustrative material, or alterations in the size of the plate. (Thompson, M. and Verner)

Statement of responsibility. A statement, transcribed from the item being described, relating to persons responsible for the intellectual or artistic content of the item, to corporate bodies from which the content emanates, or to persons or corporate bodies responsible for the performance of the content of the item. *See also* Collaborator, Compiler, Corporate body, Editor, Joint author, Mixed responsibility, Personal author, Shared responsibility. (*AACR2*)

Strip map. A map usually depicting a linear feature such as a road, waterway, or boundary, with an elongated format within which the mapped area is restricted to a narrow band. (I.C.A.)

Subordinate body. A corporate body that forms an integral part of a larger body in relation to which it holds an inferior hierarchical rank. (*AACR2*)

Subseries. A series within a series; that is, a series which always appears in conjunction with another, usually more comprehensive, series of which it forms a section. Its title may or may not be dependent on the title of the main series. (*AACR2*)

Supplement. An item, usually issued separately, that complements one already published by bringing up-to-date or otherwise continuing the original or by containing a special feature not included in the original; the supplement has a formal relationship with the original as expressed by common authorship, a common title or subtitle, and/or a stated intention to continue or supplement the original. (*AACR2*)

Supplied title. In the case of an item that has no title proper on the chief source of information or its substitute, the title provided by the cataloguer. It may be taken from elsewhere in the item itself or from a reference source, or it may be composed by the cataloguer. (*AACR2*)

Synthetic aperture radar (SAR) image. A radar in which a synthetically long apparent or effective aperture is constructed by integrating multiple returns from the same ground cell, taking advantage of the Doppler effect to produce a phase history film or tape that may be optically or digitally processed to reproduce an image. (Reeves)

Tactile (or tactual) graphics. This is a general term including anything written or drawn which employs symbols perceptible to touch. *Note:* Both "tactile" and "tactual" are in use: "tactual" has been used more frequently in the past, but "tactile" is now coming into general use. In this manual, "tactile" is used and recommended.

Tactile map. A map which employs symbols perceptible to touch.

Terrestrial globe. *See* Globe.

Terrestrial remote-sensing image. An image produced by a remote-sensing device which is located on the earth.

Tête-bêche. A tête-bêche map normally consists of either two separate works printed back to back, or a single work in two languages printed back to back. A tête-bêche atlas may be read from either end, some part of its paging being inverted relative to the rest. Such a publication will normally contain either two separate works with a title page for each work, or a single work in two languages and/or scripts (or in two versions other than language versions) with a title page in each language and/or script (or for each version). *See also* Back-to-back. (Based on: Ravilious)

Text. A term used as a general material designation to designate printed material accessible to the naked eye (e.g., a book, a pamphlet, or a broadside). (*AACR2*)

Thematic map. A map designed to demonstrate particular features or concepts. In conventional use, this term excludes topographic maps. (I.C.A.)

Title. A word, phrase, character, or group of characters, normally appearing in an item, naming the item or the work contained in it. *See also* Alternative title, Binder's title, Collective title, Cover title, Half title, Key-title, Parallel title, Running title, Spine title, Supplied title, Title proper, Uniform title. (*AACR2*)

Title block. A space on a mosaic, map, or plan devoted to identification, reference, and scale information. (Reeves)

Title page. A page at the beginning of an item bearing the title proper and usually, though not necessarily, the statement of responsibility and the data relating to publication. The leaf bearing the title page is commonly called the *title page* although properly called the *title leaf*. *See also* Added title page. (*AACR2*)

Title proper. The chief name of an item, including any alternative title but excluding parallel titles and other title information. *(AACR2)*

Topographic map. A map whose principal purpose is to portray and identify the features of the earth's surface as precisely as possible within the limitations imposed by scale. (I.C.A. mod.)

Topographic model. *See* Relief model.

Town plat. *See* Plat.

Tracing. 1. The record of the headings under which an item is represented in the catalogue.

    2. The record of the references that have been made to a name or to the title of an item that is represented in the catalogue. *(AACR2)*

    3. The facsimile transference of an image, either by drawing on to a translucent medium placed over the original, as on a light table; or by the use of transfer paper. (I.C.A.)

Transparency. A sheet of transparent material bearing an image and designed for use with an overhead projector or a light box. It may be mounted in a frame. *(AACR2)*

Triangulation diagram. A diagram which represents the stations and the observed rays of a triangulation network. (I.C.A.)

Trigonometrical diagram. *See* Triangulation diagram.

Uncontrolled photomosaic. 1. A mosaic composed of uncorrected prints, the detail of which has been matched from print to print without ground control or other orientation. (U.S. Defense Mapping Agency)

    2. (Print laydown): A photograph of an uncontrolled assembly of complete contact prints of vertical air photographs intended to serve as an index or as a map substitute. (I.C.A.)

Uniform title. 1. The particular title by which a work that has appeared under varying titles is to be identified for cataloguing purposes.

    2. A conventional collective title used to collocate publications of an author, composer, or corporate body containing several works or extracts, etc., from several works, e.g., complete works, several works in a particular literary or musical form. *(AACR2)*

Variant. A copy showing any bibliographically significant difference from one or more other copies of the same edition. The term may refer to an impression, issue, or state.

Verbal scale. The relationship which a small distance on a graphic bears to the corresponding distance on the earth, expressed as an equivalence, such as 1 inch (on the graphic) equals 1 mile (on the ground).

(U.S. Defense Mapping Agency)

Verso. 1. The left-hand page of a book, usually bearing an even page number.

    2. The side of a printed sheet intended to be read second. *(AACR2)*

See also Recto.

Vertical air photograph. An air photograph taken downwards with the axis of the camera maintained as nearly vertical as possible; the resultant photography lies approximately in a horizontal plane. (I.C.A.)

Vertical airphoto. *See* Vertical air photograph.

Vertical exaggeration. The ratio of the vertical to the horizontal scale on, for example, a relief model, plastic relief model, block diagram, profile or section. (I.C.A.)

Vertical interval. *See* Contour interval.

View. A perspective representation of the landscape in which detail is shown as if projected on to an oblique plane. *See also* Bird's-eye view, Panorama, Panoramic drawing, Perspective view, Worm's-eye view.

Volume. 1. In a bibliographic sense, a major division of a work, regardless of its designation by the publisher, distinguished from other major divisions of the same work by having its own inclusive title page,[1] half title, or portfolio title, and usually independent pagination, foliation, or signatures. This major bibliographic unit may include various title pages and/or paginations.

    2. In the material sense, all that is contained in one binding, portfolio, etc., whether as originally issued

---

1. The most general title page, half title, or cover title is the determining factor in deciding what constitutes a bibliographic volume, e.g., a reissue in one binding, with a general title page, of a work previously issued in two or more bibliographic volumes is considered to be one bibliographic volume even though the reissue includes the title pages of the original volumes.

Glossary

or as bound after issue.[2] The volume as a material unit may not coincide with the volume as a bibliographic unit. (*AACR2*)

Wall map. A map designed to be legible from a distance when it is mounted on a wall. (I.C.A.)

Wood engraving. *See* Woodcut.

Woodcut. The woodcut technique represents the transfer of ink from a raised surface to paper using direct vertical pressure. The finished print from a woodcut is conceived as black lines on a white ground. In contrast a wood engraving is conceived as white lines on a black ground. (Woodward)

Worm's-eye view. Perspective representation from a very low viewpoint. (I.C.A.)

Sources

(*AACR2*). *Anglo-American Cataloguing Rules* / prepared by American Library Association, British Library, Canadian Committee on Cataloguing, Library Association, Library of Congress ; ed. by Michael Gorman and Paul W. Winkler. — 2nd ed. — Chicago : American Library Association ; Ottawa : Canadian Library Association, 1978.

Baker, B.B., Jr., ed. *Glossary of Oceanographic Terms* / ed. by B.B. Baker, Jr., W.R. Deebel, R.D. Geisenderfer. — 2nd ed. — Washington, D.C. : U.S. Naval Oceanographic Office, 1966. Special Publication SP-35.

Boggs, Samuel W. *The Classification and Cataloging of Maps and Atlases* / by Samuel W. Boggs and Dorothy Cornwell Lewis. — New York, N.Y. : Special Libraries Association, 1945.

Evans, Frank B. "A basic glossary for archivists, manuscript curators, and records managers." *The American Archivist* 37 (1974):415–433.

Gary, Margaret, ed. *Glossary of Geology* / ed. by Margaret Gary, Robert McAfee Jr., and Carol L. Wolf. — Washington, D.C. : American Geological Institute, c1974, 1977 printing.

(I.C.A.) International Cartographic Association, Commission II. *Multilingual Dictionary of Technical Terms in Cartography.* — Wiesbaden, Germany : Franz Steiner, 1973.

Ravilious, C.P. "Tête-bêche format : a problem for descriptive cataloguing." *International Cataloguing* 8(3):35–36.

Reeves, Robert Grier, ed. *Manual of Remote Sensing.* — Sallschurch, Va. : American Society of Photogrammetry, 1975.

Stamp, Sir Dudley, ed. *A Glossary of Geographical Terms.* — 2nd ed. — New York, N.Y. : John Wiley & Sons Inc., 1966.

Steers, J.A. *An Introduction to the Study of Map Projections.* — ₍14th ed.₎. — London : University of London Press, 1965.

Stevenson, George A. *Graphic Arts Encyclopedia.* — 2nd ed. — New York, N.Y. : McGraw-Hill Book Company, 1979.

Thompson, Elizabeth H. *A.L.A. Glossary of Library Terms with a Selection of Terms in Related Fields.* — Chicago, Ill. : American Library Association, 1943.

Thompson, Morris M. *Maps for America : Cartographic Products of the U.S. Geological Survey and Others.* — Washington, D.C. : Geological Survey, ₍1979₎.

U.S. Defense Mapping Agency. Topographic Center. *Glossary of Mapping, Charting, and Geodetic Terms.* — 3rd ed. — Washington, D.C. : ₍Supt. of Documents, U.S. Government Printing Office₎, 1973.

Verner, Coolie. *The Northpart of America* / by Coolie Verner and Basil Stuart-Stubbs. — ₍Canada₎ : Academic Press Canada Limited, c1979. — p. 228.

Webster's Third New International Dictionary of the English Language, Unabridged / editor in chief Philip Babcock Gove. — Springfield, Mass. : Merriam, 1971.

Woodward, David, ed. *Five Centuries of Map Printing.* — Chicago : University of Chicago Press, 1975.

---

2. Such a composite volume bound by or for an individual owner may contain either two or more bibliographic volumes of the same work or more works published independently.

AACR	MANUAL	AACR	MANUAL	AACR	MANUAL
0.7	0.1	1.1E5	1E5	1.4C3	4C3
0.8	0.2	1.1E6	1E6	1.4C4	4C4
0.9	0.3	1.1F1	1F1	1.4C5	4C5
0.24	0.4	1.1F2	1F2	1.4C6	4C6
0.25*	0.5	1.1F3	1F3	1.4C7	4C7
0.28*	0.6	1.1F4	1F4	1.4D1	4D1
0.29	0Da	1.1F5	1F5	1.4D2	4D3
1.0A1*	0B1	1.1F6	1F6	1.4D3	4D4
1.0A2*	0B2	1.1F7	1F7	1.4D4	4D5
1.0B	0A1	1.1F9	1F9	1.4D5	4D6
1.0C*	0C1	1.1F10	1F10	1.4D6	4D7
1.0D	0Db	1.1F11	1F11	1.4D7	4D8
1.0D1	0D1	1.1F12	1F12	1.4E1	4E1
1.0D2	0D2	1.1F13	1F13	1.4F1	4F1
1.0D3	0D3	1.1F14	1F14	1.4F2	4F2
1.0E	0E	1.1F15	1F15	1.4F3	4F3
1.0F	0F1	1.1G1	1G2	1.4F4	4F4
1.0G	0G1	1.1G2	1G3	1.4F5	4F5
1.0H*	0J1	1.1G4	1G5	1.4F6	4F6
1.1A2	1A2	1.2A2	2A2	1.4F7	4F7
1.1B1	1B1	1.2B1	2B1	1.4F8	4F8
1.1B2	1B2	1.2B2	2B2	1.4G	4G
1.1B3	1B3	1.2B3	2B3	1.4G2	4G2
1.1B4	1B4	1.2B4	2B4	1.4G3	4G3
1.1B5	1B5	1.2B5	2B5	1.5A2	5A2
1.1B6	1B6	1.2C1*	2C1	1.5A3*	5A3
1.1B7*	1B7	1.2C2	2C2	1.5B5*	5B28
1.1B8*	1B8	1.2D1	2D2	1.5E1	5E1
1.1B9	1B9	1.2E1	2E1	1.6	6
1.1C1	1C1	1.4A2	4A2	1.6A2	6A2
1.1C2	1C2	1.4B1	4B1	1.6B	6B1
1.1C3	1C3	1.4B2	4B2	1.6B1	6B1
1.1D1	1D1	1.4B3	4B3	1.6B2	6B3
1.1D2	1D2	1.4B4	4B4	1.6C	6C
1.1D3	1D3	1.4B5	4B5	1.6C1	6C1
1.1D4	1D4	1.4B6	4B6	1.6D	6D
1.1E1	1E1	1.4B7	4B7	1.6D1	6D1
1.1E2	1E2	1.4B8	4B8	1.6E	6E
1.1E3	1E3	1.4C1	4C1	1.6E1	6E1
1.1E4	1E4	1.4C2	4C2	1.6F	6F

*These rules, or parts thereof, have been modified in the Manual.

Concordance

AACR	MANUAL	AACR	MANUAL	AACR	MANUAL
1.6F1	6F1	2.5B16	5B19	3.1C2	1C4
1.6G	6G	2.5B17*	5B20	3.1D	1D
1.6G1	6G1	2.5B18	5B21	3.1E	1E
1.6G2	6G2	2.5B19	5B22	3.1E2	1E6
1.6G3	6G3	2.5B20	5B23	3.1F	1F
1.6H	6H	2.5B21	5B24	3.1F2	1F8
1.6H1	6H1	2.5B22	5B25	3.1G	1G
1.6H2	6H2	2.5B23*	5B26	3.1G1	1G1
1.6H3	6H3	2.5C1*	5C2b	3.1G3	1G4
1.6H4	6H4	2.5C2*	5C2c	3.1G4	1G6
1.6H5	6H5	2.5C3*	5C3	3.1G5	1G7
1.6J	6J	2.5C4	5C2d	3.2	2
1.6J1	6J1	2.5C5	5C2e	3.2A	2A
1.7A2	7A2	2.5C7	5C2f	3.2A1	2A1
1.7A3	7A3	2.5C8	5C2g	3.2B	2B
1.7A4	7A4	2.5D1	5D2	3.2B1	2B1
1.7A5*	7A5	2.5D2	5D2	3.2B3	2B4
1.7B15	7B15	2.5D3	5D2	3.2B4	2B5
1.7B16	7B16	2.5D5	5D2	3.2B5	2B7
1.8A2	8A2	2.7B13*	7B13	3.2C	2C
1.8B1	8B1	2.13*	0B5	3.2C1*	2C1
1.8B2	8B2	2.14A*	1B12	3.2D	2D
1.8B3	8B3	2.14B	1B13	3.2D1	2D1
1.8B4	8B4	2.14C*	1B14	3.2E	2E
1.8C1	8C1	2.14D	1B15	3.3	3
1.8D1	8D1	2.14E*	1B16	3.3A	3A
1.8E1	8E1	2.15A*	2B9	3.3A1	3A1
1.8E2	8E2	2.15B	2B8	3.3A2	3A2
1.9A	9A	2.16A	4B9	3.3B	3B
1.9B	9B	2.16B	4C8	3.3B1*	3B1
1.10A*	10A	2.16C	4C9	3.3B2	3B2
1.10B	10B	2.16D*	4C10	3.3B3	3B3
1.10C	10C	2.16E	4C11	3.3B4	3B4
1.10C1	10C1	2.16F	4D9	3.3B5*	3B5
1.10C2	10C2	2.16G	4D10	3.3B6*	3B6
1.10C3	10C3	2.16H*	4F10	3.3B7	3B7
1.10D	10D	2.16J	4F11	3.3B8	3B8
1.11A*	11A	2.16K*	4G5	3.3C	3C
1.11B	11B	2.18C*	7B15	3.3C1	3C1
1.11C	11C	2.18D*	7B10b	3.3C2	3C2
1.11D	11D	2.18F*	7B20b	3.3D	3D
1.11E	11E			3.3D1*	3D1
1.11F	11F	3.0	0	3.3D2*	3D2
		3.0A	0A	3.4	4
		3.0B	0B	3.4A	4A
2.0B1*	0B4	3.0B2	0B3	3.4A1	4A1
2.0B2*	0B7	3.0B3	0B6	3.4B	4B
2.1B2*	1B11	3.0C	0C	3.4B2	4B9
2.5B1	5B4	3.0D	0Dc	3.4C	4C
2.5B2	5B5	3.0E	0E	3.4D	4D
2.5B3	5B6	3.0F	0F	3.4D1	4D2
2.5B4	5B7	3.0G	0G	3.4E	4E
2.5B5	5B8	3.0H	0J	3.4F	4F
2.5B6	5B9	3.0J*	0K	3.4G1	4G1
2.5B7	5B10	3.1	1	3.4G2	4G4
2.5B8	5B11	3.1A	1A	3.5	5
2.5B9	5B12	3.1A1	1A1	3.5A	5A
2.5B10	5B13	3.1B	1B	3.5A1	5A1
2.5B11	5B14	3.1B2	1B10	3.5B	5B
2.5B12	5B15	3.1B3*	1B8	3.5B1*	5B1
2.5B13*	5B16	3.1B4*	1B7	3.5B2*	5B2
2.5B14	5B17	3.1C	1C	3.5B3	5B3
2.5B15*	5B18				

AACR	MANUAL	AACR	MANUAL	AACR	MANUAL
3.5B4*	5B27	3.8C	8C	13.5B	13E2
3.5C	5C	3.8D	8D	13.6	13F
3.5C1	5C1	3.8E	8E		
3.5C2	5C2a	3.9	9	21.1A1	A.1
3.5C3	5C3	3.10	10		
3.5C4*	5C4	3.11	11	B.1*	H.1
3.5C5	5C5			B.4	H.2
3.5D	5D	4.0B2*	0B8	B.5	H.3
3.5D1*	5D1	4.1F2	1F16	B.6	H.4
3.5D2	5D2	4.1F3	1F17	B.7	H.5
3.5D3	5D3	4.4B1*	4F9	B.8	H.6
3.5D4	5D4	4.5B1	5B29	B.9*	H.7
3.5D5	5D5	4.7B1*	7B1b–7B1g	B.13	H.8
3.5E	5E	4.7B7*	7B7b	B.14	H.9
3.6	6	4.7B8*	7B9	B.15	H.10
3.6A	6A	4.7B9*	7B7c		
3.6A1	6A1	4.7B14*	7B14	C.0	J.0
3.6B	6B	4.7B15*	7B15	C.1	J.1
3.7*	7	4.7B22*	7B20c	C.1A	J.1A
3.7A	7A			C.1B	J.1B
3.7A1	7A1	8.5C12*	5C3	C.1C	J.1C
3.7B	7B	8.5C16*	5C3	C.1D	J.1D
3.7B1	7B1	8.5D4*	5D6	C.1E	J.1E
3.7B2	7B2	8.5D5*	5D7	C.2	J.2
3.7B3*	7B3	8.7B8*	7B7d	C.2A	J.2A
3.7B4	7B4			C.2B	J.2B
3.7B5	7B5	11.3A*	3D3	C.2C	J.2C
3.7B6	7B6	11.5B1	5B30	C.3	J.3
3.7B7	7B7	11.5C1	5C6	C.4	J.4
3.7B8	7B8	11.5D3*	5D8	C.4A	J.4A
3.7B9	7B9	11.7B10	7B10c, d	C.4B	J.4B
3.7B10*	7B10a			C.4C	J.4C
3.7B11	7B11	12.1B2*	6B2	C.4D	J.4D
3.7B12	7B12	12.2B3*	2B6	C.4E	J.4E
3.7B14	7B14	12.6B1*	6G2	C.5	J.5
3.7B18*	7B18			C.5A	J.5A
3.7B19	7B19	13	13	C.6	J.6
3.7B20	7B20	13.1	13A	C.7	J.7
3.7B21	7B21	13.2*	13B	C.7A	J.7A
3.8	8	13.3	13C	C.7B	J.7B
3.8A	8A	13.4	13D	C.7C	J.7C
3.8A1	8A1	13.5	13E	C.7D	J.7D
3.8B	8B	13.5A*	13E1		

INDEX

The index covers the rules (including applications to the rules) and appendices, but not examples or works cited in any of the rules or appendices. All index entries refer to rule numbers. The applications are included in the rule numbers, i.e., they are not identified separately. *Glossary* indicates that a term is defined in the Glossary.

As the rules are based upon bibliographic conditions rather than specific cases, kinds of work have been indexed only when actually named in a rule. There is no entry under *Encyclopedias, Directories,* etc., because they could represent several bibliographic conditions.

The index is arranged according to *ALA Rules for Filing Catalog Cards* / prepared by the ALA Editorial Committee's Subcommittee on the ALA Rules for Filing Catalog Cards ; Pauline A. Seeley, chairman and editor. — 2nd ed. — Chicago : American Library Association, 1968.

Abbreviations used in the index are: App. (Appendix) and n (footnote).

Index

Index

Index

Volumes
  definitions, Glossary
  number of (atlases), 5B3, 5B11, 5B20– 5B25,
    5B28

Wall maps, definition, Glossary
Width, 5D
  atlases, 5D2
  see also Dimensions
"With" notes, 7B21
Wood engravings, *see* Woodcuts
Woodcuts, definition, Glossary

Words or phrases
  added to statements of responsibility, 1F8
  indicating function of distributor, etc., 4D4
  indicating publishing, etc., 4D
  indicating responsibility, 1F1, 1F12
Worm's-eye views, definition, Glossary
Writers, *see* Personal authors; Personal names, as
  title proper
Writing, place of (manuscript), 7B9

Year(s), *see* Date(s)

Zones (celestial charts), *see* Declination zones

Designed by Vladimir Reichl

Composed by Automated Office Systems Inc.
    in Times Roman on a Text Ed/VIP
    phototypesetting system

Text printed on 50-pound Antique
    Glatfelter, a pH-neutral stock, by
    Chicago Press Corporation.
    Four-color insert printed by LithoGraphics.

Cover printed on Joanna B-grade cloth
    by Chicago Press Corporation.
    Bound by Zonne Bookbinders.

MAPS 162/84